D1341793

WEATHER FORECASTING
FOR AGRICULTURE
AND INDUSTRY

WEATHER FORECASTING FOR AGRICULTURE AND INDUSTRY

A SYMPOSIUM

EDITED BY

JAMES A. TAYLOR

Reader in Geography
University College of Wales, Aberystwyth

CONTRIBUTORS

C. V. Smith
H. H. Lamb
J. P. Hudson
A. N. Duckham
W. H. Hogg
W. L. Thomas
Miss M. G. Roy and J. M. Peacock
R. A. Buchanan
A. H. Perry
A. H. Paul
J. D. McQuigg
J. S. Hay and C. P. Young
R. C. Goodhew and E. Jackson
R. S. Gander
J. K. Page
C. V. Barnett

Rutherford · Madison · Teaneck
FAIRLEIGH DICKINSON UNIVERSITY PRESS

96064

First American edition published 1972 by
Associated University Presses Inc
Cranbury New Jersey 08512

Library of Congress Catalogue Card Number 72–6550
ISBN 0–8386–1266–0

Printed in Great Britain by Bell & Bain Ltd, Glasgow.

List of Contents

List of Illustrations

Foreword

SIR GRAHAM SUTTON, CBE, DSc, FRS
*Vice-President of the University College of Wales, Aberystwyth; formerly
Director-General of the Meteorological Office*

I am very glad to respond to the invitation of the editor to write a foreword to this volume. The earliest form of weather forecasting, by signs and portents, which has come down to us in proverbial form as 'weather lore', is inextricably linked with husbandry. Climatology came much later, beginning with monkish records and country diaries, and it was some time before it was realised that the variations of weather and climate are far more complex than the motions of the stars that so dominated medieval thought. Today we are living in the period of the most rapid advance yet achieved in the science of the atmosphere; since about the middle of this century the forecasting of weather over short periods has become increasingly precise and accurate, thanks to a better understanding of dynamical meteorology and the invention of the high-speed computer, and even long-range weather forecasting (or, more accurately, the prediction of climatic anomalies for periods up to a month ahead) has made sufficient progress to justify publication on a routine basis.

The primary need of the crop farmer, to know well in advance what, where, and when to sow and reap, remains. The sciences to which he must look for an answer, agriculture and economics, geography and meteorology, can help in differing degree but undoubtedly the most uncertain advice is that from the meteorologists. Ideally, the farmer would like to know something about weather prospects a year in advance, but at present and probably for a long time to come, this is beyond meteorology. It is now becoming increasingly clear that the predictability of the behaviour of weather systems by mathematical methods, starting from a known situation, is limited, the period of validity of the forecast being very dependent on the spatial scales of the phenomena studied. Even the largest-scale features, the motions of the planetary waves that determine, eg, whether we have a 'good' or a 'bad' summer, do not seem capable of prediction by dynamical methods for more than a few weeks ahead. If climatic variation is not entirely a random phenomenon we must search past climatic records, however derived, for guides to the future and that is one reason why so much attention is now being given to the application of highly sophisticated statistical techniques to climatology.

The contributions to the present volume show a fascinating variety of approach to these important matters. They represent the essential first steps towards the definition, if not the solution, of problems of increasing significance in a hungry world. It is appropriate that Aberystwyth should take the lead in this field, for its record in agricultural science and geography is outstanding. But as will be seen from the list of the papers read at the symposium it is not only the farmer who has a need for meteorological advice; industry also is becoming increasingly weather-conscious. This record of a highly successful meeting should therefore appeal to a wide circle of readers.

Preface

The Objective

This volume is devoted to a specialised aspect of applied meteorology—the value and use of weather forecasts in agriculture and industry, with particular reference to Britain. It is a significant fact that the number of inquiries seeking weather advice from our Meteorological Office has doubled in the past decade. These inquiries come from a wide variety of sources, including industrialists, agriculturalists, and the general public. This spectacular acceleration of consumer-demand will no doubt be maintained in the future. Are our weather services equipped to meet this new challenge? How could our current public weather forecasting services be improved so that the forecasts themselves are more precise and more accurate and also acquire a style and format more immediately intelligible to the consumer? How can the consumer himself be educated to a more intelligent appreciation of weather forecasts as applied to both social and economic activities?

In an increasingly competitive and cost-conscious world, the financial impact not only of weather hazards but also of cumulative wear-and-tear under weather stress is at last being fully understood. Agriculture and numerous industries which operate out-of-doors are directly weather-sensitive. Many 'housed' industries use weather-sensitive processes or materials. Certain basic industries, eg, the power industries, transportation, etc, are continuously vulnerable to weather variations, and this has repercussions throughout a wide range of industrial—and social—activity.

Weather dependence may be instant or cumulative, short-term or long-term. Decision-making in day-to-day operations in agriculture and the construction industries; the marketing distribution of weather-sensitive stocks, eg, of certain foodstuffs and beverages (see Gander, Chapter 15 herein); the fluctuating demand for gas and electricity due to the weather and the need for adequate but economic storage of power in the event of a sudden cold spell, for instance; success and choice of alternatives in leisure activity, especially tourism as conditioned often by recent individual experiences (see Taylor, Chapter 1 herein)—these are but a few examples of the range of dependence.

Perhaps one particularly interesting and topical case-study—that of the British motorways—might be taken in a little detail to set the scene and indicate the objective.

Our motorways crystallise, and in many ways caricature, the general problem. Since they have been introduced to provide swift communication by road between our major conurbations and industrial areas, relatively straight and, so far as possible, gradient-free routes have been selected. Some weather hazards have thereby been accentuated, particularly from the driving point of view. First of all, they are open to wind exposure which may vary suddenly between embanked and unembanked sections. Large vehicles with a high centre of gravity are particularly vulnerable, but variable cross winds are a danger to all vehicles travelling at speed. Second, they happen to have been located in several largely built-up lowland areas which are inherently liable to persistent fog, eg, south Lancashire, parts of Yorkshire and the Midlands generally. However, it is the occurrence of sudden fog-banks which is the worst cause of multiple pile-ups on motorways. Whilst many of these fog-banks are sporadic in occurrence and more or less impossible to forecast, some sections of motorways are more liable to fog-banks either by virtue of the motorway's location in a hollow near a water course or because of its design. The Thelwall viaduct on the M6, on the Lancashire/Cheshire border, is a notorious example. Third, the very extent of the motorway surface and the differential wear and tear of fast and slow lanes, means many small-scale variations in the micro-climatological conditions of the road surface. Presence or absence of surface water, variable dryness and moistness of road surface, actual road-surface temperature, all directly affect vehicle performance and control, especially if they vary suddenly. Motorway traffic then is liable to encounter these weather hazards at fast speeds. Safety margins are reduced and terrible multiple accidents may occur. The solutions to the problem lie (a) in the education of motorway drivers to the full meaning of the motorway environment, (b) in the establishment of efficient local weather forecasting and hazard-warning systems, eg, against ice (see Chapter 13 by Hay and Young herein), fog or floods, and (c) the provision of warning systems and safety devices on the vehicles themselves. However, although the cost in lives and materials of motorway accidents is to be deplored, it should be emphasised that other roads, taken cumulatively, provide even more deplorable evidence. In cost-benefit terms the motorways save an enormous amount of economic and social time, but at what stage will they recover the large sums involved in their initial construction? Those large sums, incidentally, include an invisible, variable cost, viz, work-days (for man and machine) lost due to bad weather. It appears that a crude standard clause is included in most British engineering contracts to allow for this. The resultant variations per contract between estimates and expenditure on this count must be very large. Ironically, the Meteorological Office has information available on probable work-day loss, which is apparently not used. The engineering companies take the view, quite rightly, that weather losses are one of a number of managerial losses incurred during construction of the motorway but then assert, quite wrongly, that weather-loss could not be extracted from their work-diaries, which in any event they will not divulge. The situation in the USA is somewhat better and certainly more co-operative and economic for all parties concerned (see McQuigg, Chapter 12 herein).

The motorways, then, mirror the current stage reached in development and application of our weather forecasting technology. An intensifying system is being more weather-sensitive. Severe weather hazards and bad accidents are occurring more frequently so both the authorities and the general public are becoming more weather-conscious. The economic and social demand for better weather services and weather communication precipitates ice and fog warning systems. It is now up to all motorway drivers, the consumers, to become educated to the new services as they become available. However, the lack of more specific accounting in motorway contracts in Britain for probable work-day loss due to the weather is unforgivable on all economic grounds.

Our weather forecasting services are broadly in a similar stage of development. An intensifying economy and society becomes more cost-conscious and more comfort-demanding. More intelligent adjustment of programmes and materials to the expected weather conditions is required. More specific short- and long-term weather forecasts are needed not only to save money but also to reduce the risk of social inconvenience.

The abstraction of the weather factor in cost-benefit terms leads to a re-valuation of our weather services in both social and economic terms. It follows that the long-term planning of the built environment and regional plans as such should take full account of the local climatological changes which may be induced and also of the possible effects generally of macro-climatic changes which are now better understood (see Lamb, Chapter 3 and Page, Chapter 16, herein).

It behoves us to conduct detailed consumer research, improve the weather forecast, tailor it to the needs of the user, and communicate it effectually. Our blueprints for built-up areas and regional environments of the future should take full cognizance of the local climatic differences the artificially created landscapes will incorporate as well as the possible regional impact of known macro-climatic changes. It is towards this general objective that this volume of essays is pointed.

The Contents
Since 1958 annual symposia have been convened at the University College of Wales, Aberystwyth, on the broad theme of agricultural climatology. The fourteenth in the series was held on 17–18 March 1971 and was devoted to the subject of weather forecasting as related to agriculture and industry. About one hundred delegates attended, including representatives from the universities, the research stations, the advisory services, the scientific civil service, local authorities, agriculture, industry, administration, and the communication media. The interdisciplinary traditions of the symposia series were thus fully maintained and the diversity of viewpoints available, both academic and technical, guaranteed full and critical discussion. Seventeen papers were presented, three dealing with the problems and policies of weather forecasting, five related to agricultural activities, and nine discussing a variety of industrial applications. They are published herewith in that sequence, incorporating as appropriate some of the major points derived from the lively discussions which were generated at the symposium.

Weather forecasting is sometimes regarded as an art rather than a science. The analogue technique of comparing the current weather situation with previous similar ones is familiar but can be quite subjective and liable to error. On the other hand the availability of the best computers for the assimilation and processing of weather data will not necessarily ever lead to a substantial improvement in the present accuracy of the daily weather forecasts, although particular aspects of these forecasts, eg, rainfall amounts, will be more quantified and more precise in the future. What is more crucial at present (as emphasised above) is to orientate the forecasting service towards consumer needs and provide specific weather information for particular economic or social ends. The editor in Chapter 1 envisages the relationships between forecaster and consumer in terms of the flow of weather information, public and private, as inputs into entrepreneurial systems. Vital to this flow are the communication links, the controls imposed by the mass media of television and radio, and the need for *inter*-communication at all stages of the system to achieve administrative and economic efficiency. It is a type of weather technology which Mr C. V. Smith (Meteorological Office) is advocating in Chapter 2. Meteorological marketing is just as important, if not more important, than the way the forecast is derived. This point of view implies a significant degree of re-organisation of current priorities in our meteorological services. Mr H. H. Lamb (Meteorological Office) in Chapter 3 discusses briefly some of the problems and practices of longer-range weather forecasting. As an international authority on the subject of climatic change, he has been able to apply his findings on recent trends in climate to assessing the medium and long-term climatic probabilities. The now published monthly forecasts are evidence of increasing confidence in this sphere.

Five chapters of agricultural implication follow. In Chapter 4 Professor J. P. Hudson (Director of the Research Station at Long Ashton, Bristol) reminds us very forcibly that the land-use and agricultural plans for twenty-five or fifty years ahead, including the work of the geneticist in breeding new plants and animals, must take into account the projected climatic conditions of our future environments. Other competitive forces are involved, economic, social, and political, but the environmental trends will be equally relevant on their own terms, and, if they can be predicted, ecological and economic benefits will accrue. Emeritus Professor A. N. Duckham (sometime Professor of Agriculture in the University of Reading), another international authority on the impingement of agro-meteorological problems on farm management and food supply, examines the role of weather forecasting in agricultural management and decision-making in Chapter 5. There follows an interesting report (Chapter 6) by Mr W. H. Hogg (sometime Senior Meteorological Officer based at Bristol) on the actual weather forecasting requirements of specific types of agriculture in Wales, the West Midlands, and the South-West. The analysis is based on over one hundred and fifty questionnaires filled in voluntarily by farmers and growers in those regions; this was a 60 per cent return on the original circulation. Improvements in the weather forecasting services available to farmers are suggested, eg, the re-timing of certain broadcasts, the tailoring of forecasts to special agricultural tasks.

Mr W. L. Thomas, a practising farmer from West Pembrokeshire, had already achieved literary and technological fame in publishing a book (in co-authorship with Mr P. W. Eyre) in 1951 (Faber and Faber) on the subject of *Early Potatoes*. The editor was fortunate enough to be able to track him down at Clynderwen where he still grows early potatoes most successfully. His contribution, written from the practical growers' viewpoint, is a masterly demonstration of the way in which skilful management can reduce and even eliminate weather risks, and keep down production costs at the same time. His own medium- and long-term seasonal forecasting scheme is ingenious and proof of its success is evident from his regularly high profit margins. Moreover, he is quite content to inform all willing competitors about his methods of production! The co-authors of the next chapter, Miss M. G. Roy (seconded from the Meteorological Office to the Grassland Research Institute, Hurley) and Mr J. M. Peacock (of the same Research Institute) demonstrate a model for the seasonal forecasting of spring growth and the flowering of forage crops in the British Isles. It is suggested that under certain circumstances a forecast of soil temperature could be used to predict grass yields. Methods of predicting the date of heading from mean soil temperatures are also discussed. This ends the specifically agricultural section.

The next nine chapters (9 to 17 inclusive) are concerned with weather forecasting for industry. First of all, Mr R. O. Buchanan (Meteorological Office) discusses the topic in general and shows how the demand for meteorological services has more than doubled in the last decade. Special studies are made of the construction, manufacturing, and service industries subsequent to an outline of the forecasting services provided by the Meteorological Office to meet the general public interest in weather in the United Kingdom. The restrictive effects of the communication media are fully pin-pointed. Direct contact with the enterprising consumer is so much more efficient than a general system of weather forecasting. Future excessive demands for weather advice might well be met by private consultants and by meteorological specialists appointed within specific companies.

Dr A. H. Perry, Lecturer in Geography at University College, Swansea, attempts in Chapter 10 to relate weather trends to the expansion since the 1940s of the ski-resorts in the Cairngorms in Scotland and concludes that the number and type of climatic problems facing the industry would encourage the investment of smaller capital outlays in projects which are more flexible to operate according to weather and season. Continuing the theme, Dr A. H. Paul of the Geography Department, University of Saskatchewan, Canada, studies in Chapter 11 the effects of weather on daily attendance at three of the outdoor recreation areas of Canada. His work reveals the large number of variables involved and the difficulty of identifying the precise combination of meteorological parameters which affect a given leisure activity at a particular stage. This subject has special significance in view of the increasing national and international pressures on leisure resources.

Dr J. D. McQuigg, Research Meteorologist at the University of Missouri, Columbia, USA offers, in Chapter 12, valuable simulation model studies of the impact of weather factors on road construction and the movement of heavy equipment in agricultural operations. This work is most pertinent, for example, to motorway construction in this country. The implications for large-scale mechanised agriculture are equally enormous. In Chapter 13 Dr J. Hay and Dr C. P. Young of the Road Research Laboratory deal with the problem of improving weather forecasting techniques to reduce the amounts, and therefore the cost, of salt to be spread on roads to prevent icing. The national cumulative cost of salt used for this purpose is about £7 million but it is also estimated that the cost of corrosion of vehicles (which is accelerated by salt forays) is of the order of £60 million a year. The economic case is clear. Experimental trials with ice warning systems on a stretch of the M62 Trans-Pennine Motorway and the M4 from London and South Wales are imminent. The reduction of accident rates and the maintenance of communications would be further benefits from this research if extended.

Water is now recognised as a vital, if not *the* vital, national commodity. Hence the supreme importance of hydrological forecasting. Mr R. C. Goodhew of the Severn River Authority and Mr E. Jackson of the Dee and Clwyd River Authority contribute Chapter 14 on the subject of weather forecasting and river management. The dependence of hydrological forecasting on accurate *local* forecasting is stressed. The better the forecast, the

B

lower the cost of water supplies, and the less frequent is flood damage. The contributions of telemetry, radar, and catchment models are discussed in relation to developments in weather forecasting techniques.

In Chapter 15 Mr R. S. Gander, an operational scientist attached to one of the large brewery companies, examines the role of weather factors in fluctuations in the demand for beer. The analysis applies theoretical econometric techniques to economically available historical data. It is estimated that lack of operational adjustment to weather can waste £3·5 million per annum of the national retail trade in beer alone. Extrapolate this scale of costs to other beverages and foodstuffs which are weather-sensitive and formidable losses would emerge.

Planning our future built environment should incorporate the prediction of the urban climate as a modification of the better known 'green-field' site. Professor J. K. Page, of the Department of Building Science in the University of Sheffield, discusses problems in this type of prediction work in Chapter 16. The lack of suitable techniques for urban climatological forecasting is admitted but it is hoped to stimulate further studies of the net radiation balance of towns, the pollution climate, and the vertical as well as horizontal climates of built-up areas. So many of our conventional meteorological stations avoid urban areas (as well as high ground) to escape the very anomalies the urban climatologist requires to study.

Finally, in Chapter 17, Mr C. V. Barnett of the Central Electricity Generating Board discusses the immediate effects of the weather on the short-term forecasting of electricity demand. Electricity cannot be stored; it must be used as it is produced. Since it takes three hours or more to put an additional large generator on load, very accurate and promptly available weather forecasts, eg, of sharp falls of temperature or rapid reduction of visibility, are absolutely essential. Thus the smooth and economic operation of the National Grid System leans heavily on the weather services. Methods are described which normally result in estimates accurate to within 1·5 to 2·0 per cent of actual demand.

Editor's Acknowledgements

The editor has pleasure in acknowledging the assistance of many in the preparation of this volume and the symposium which created it. For placing certain secretarial, cartographic and financial services at my disposal, I thank Professor C. Kidson, BA, PhD, Head of the Department of Geography at Aberystwyth. The Director of the Welsh Plant Breeding Station, Professor P. T. Thomas, PhD, CBE, kindly gave permission to hold the symposium at Plas Gogerddan and thanks are due to the Secretary, Mr J. W. Ellis, DRA, ACIS, for arranging services and hospitality at the Station. I am grateful to the University College of Wales for support and to the Vice-Principal, Professor Ivor Gowan, BA, MA, for a grant in aid of hospitality. I acknowledge the encouragement and help of my colleagues and students of the Department of Geography at Aberystwyth. A special thank-you is due to Mr David Unwin and Mr John Harrison for direct assistance at the symposium itself. Dr John Cooper of the Welsh Plant Breeding Station very kindly and efficiently took the chair for some of the papers.

The success of the symposium was mainly due to the prompt and effective co-operation of the speakers and delegates. A special vote of thanks goes to my past Research Assistant, Mrs Cheryl Fitzgibbon, for helping to organise the meeting at such short notice but so efficiently, and to my current Research Assistant, Miss Janet Davies, who has helped with editorial work and prepared the indexes. Miss Carol Parry deserves special commendation for abstracting the tapes and for coping so well with the burden of typing. She was assisted with equal conscientiousness by Miss Linda James. The cartography and photography sections of the Department, in the charge of Mr Morlais Hughes and Mr. David Griffiths, respectively, provided excellent supporting services as usual. Mr Michael Gelli Jones deserves special mention since he drew the majority of the maps and diagrams.

Finally, I should like to acknowledge the continued co-operation and wise council always made freely available by Mr William H. Hogg, MA, who has been a tower of strength to me in organising and maintaining the continuity of the Aberystwyth symposia since they began in 1958.

Aberystwyth
19 November, 1971   J. A. Taylor

CHAPTER ONE   J. A. TAYLOR

The Revaluation of Weather Forecasts

Pure and Applied Meteorology

Meteorology is fundamentally the scientific study of the atmosphere and its processes and has achieved notable and accelerated progress in recent decades. The new technology based on upper air sensing, the monitoring of data by satellite, and rapid calculations by computer, has boosted developments to a spectacular scientific level. Much of the intellectual output, however, may be properly regarded as 'pure meteorology'; it is largely the work of physicists, and the results are expressed in mathematical terms.

In contrast, what may be designated 'applied meteorology' has languished. This is concerned with interactions between the atmosphere and physical and biological processes in the crop and animal environment often *below* the Stevenson Screen level to which so many meteorological and climatological analyses are referred for reasons of standardisation. It is becoming increasingly concerned also with the interactions between the atmosphere and man as a decision-maker, and with the inevitable economic and social repercussions (Maunder, 1971). It follows that applied meteorology has a contribution to make to the social sciences just as pure meteorology enables the establishment and testing of physical principles within the realms of the pure and experimental sciences. The concerted and penetrating efforts achieved in pure meteorology serve only to expose the contemporary neglect and underdevelopment of applied meteorology.

The Market for Weather Forecasts

One of the eventual outputs of primary meteorological inquiry is the weather forecast which is the meteorological package most frequently placed at the disposal of man in his economic and social pursuits, some of which may be more weather-sensitive than others.

The increasing scale and intensity of social mobility, for both work and leisure, has sharpened the interest in, and demand for, weather forecasts. At the same time continual pressure to increase economic productivity within the limits of available resources, be they space, soils or capital, etc, has created a new perception of both incidental and cumulative costs or losses due either directly or indirectly to the weather.

It is accepted that British weather is among the most variable and most unpredictable in the world. But, even allowing for this, one might justifiably ask whether the standards achieved in British weather forecasting bear any relationship to the success of the rapidly advancing research frontier in pure meteorology referred to above. Can forecasts be improved and refined for their own sake and also to meet the rapidly expanding markets for weather advice and information which must certainly emerge in the future? Is it possible to orientate forecasts to specific rather than general economic and social needs? What role is being played by the communication media, both official and unofficial, and how best can forecasts be conveyed to the individual consumer who may well require to be educated to their value and application? The answers to these questions will emerge more freely by reference to the conceptual model depicted in Fig. 1.1.

The Flow of Weather Information (Fig. 1.1)
The contribution of pure meteorology (1a) is translated into information (2) available via the Meteorological Office. The remainder of the flow diagram is concerned to illustrate what is essentially the field of applied meteorology, defined broadly as the study of the relationships between climate and weather factors and economic and social activities. The flow diagram is largely self-explanatory but several features require emphasis. Of the two interfaces marked, A–A represents the *communication* of weather information to the entrepreneur, and the B–B interface represents the cumulative and instant effects of climate and weather *directly* on industry or agriculture. It is with interface A–A that this book is primarily concerned. The role of the communication media (7) is crucial at A–A just as the role of physical and monetary weather-proofing (12) is crucial at B–B. At all times, however, the universal key variable is the entrepreneur himself (10) who is not only potentially accessible to both indirect and direct, official and unofficial

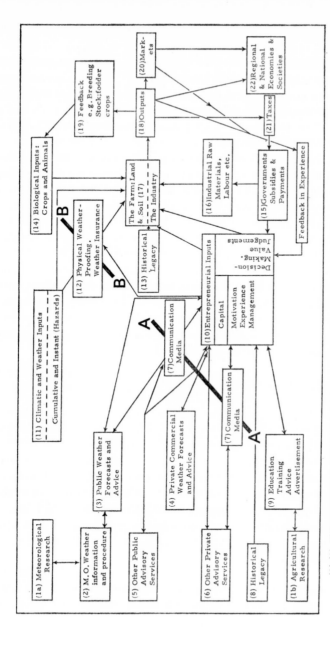

Fig. 1.1 The origin and flow of inputs of weather information within producing systems, illustrating two major interfaces

(A)-(A) THE COMMUNICATION OF WEATHER INFORMATION TO THE ENTREPRENEUR
(B)-(B) THE CUMULATIVE AND INSTANT EFFECTS OF CLIMATE AND WEATHER DIRECTLY ON THE PRODUCING SYSTEM

weather advice ((1) to (9) inclusive) but also may be variably impervious to such information sources. His decision-making and value judgements are controlled by his available capital, his motivation, his experience (including possibly some historical legacy (8)) and his management skills. At least, the way he manages his resources (17), his climatic environment (11), and his available and/or selected information sources, will be key determinants of his outputs (18) within the politico-economic framework of the time. Thus the production process which is more or less weather-sensitive is conceived as a series of linkages in which the following stages are vital:

(a)   the preparation of public weather forecasts for general consumption (3) to (10).

(b)   the availability on request to the entrepreneur of specific weather information for personal consumption (10) to (3).

(c)   the standards and efficiency of the communication media, viz, television, radio, newspapers, official Meteorological Office statements and publications (7).

(d)   the two-way potential access between private commercial forecasting and the individual consumer (4) and (10).

(e)   the role of the entrepreneur in introducing physical and/or monetary weather-proofing (12) on his own initiative (10) or on the receipt of public (3) or private (4) weather advice.

(f)   the education of the entrepreneur to the availability and optimum use of weather advice (9 and 10) or to the management finesse of weather problems (10 and 17).

(g)   the impingement of types of available weather advice on management decision-making, including the effects of intuitive value judgements based on local weather knowledge or weather lore.

(h)   the effects of feedback in the light of 'weather experiences'.

(i)   the entrepreneurial outputs (18) and the regional and national economies and societies (22) as related in particular to cost/benefit arguments.

The above stages will now be considered as applied to Britain, citing case-studies as appropriate and indicating paths to improvements.

Preparation of Public Weather Forecasts
The daily weather forecasts *as currently published* scarcely justify the image of the Meteorological Office as a high-powered research machine. At the time of writing (1971), the installation of bigger and better computers promises forecasts of rainfall *amounts* but the crucial problem of estimating more precisely the timing of onset and cessation of rainfall in particular districts remains. The major persistent weakness of forecasts is their *generality* (Gordon and Bestwick, 1969). Some attempt is made to express forecasts in *regional* terms but the 'regions' adopted do not necessarily possess any distinctive homogeneity of local weather and climate. In fact the diversity of altitude and exposure within the designated areas may vastly reduce the local applicability of the regional forecast statement. Again, since the forecasts are concerned with stating the weather *probabilities* for certain areas over certain time periods, the phraseology used is vague rather than precise, flexible rather than rigid. Such expressions as 'rain at times' and 'bright periods and showers' appear quite regularly and although they may be meaningful to the scale of the forecast statement they are meaningless to particular localities at particular times. The monthly weather forecasts, although published, are still regarded as experimental by the Meteorological Office. They have become more consistently satisfactory this year (1971) as compared with last, with the unforgettable exception of June 1971 when prolonged bad weather followed a forecast of substantial periods of good weather—all this occurring at the start of a tourist and 'outdoor' month when the public are more sensitive than usual to weather variations and weather forecasts.

The daily, and eventually the monthly, forecasts should be more particularised along the lines of forecasts already currently available on telephoning the nearest station of the Meteorological Office—lists of the appropriate telephone numbers are given in the directories. Similarly, a personal inquiry to the Office is a means of obtaining a particularised forecast for a specific location at a specific time. The currently published forecast statements are merely a pale reflection of the enormous data source from which they emanate. The forecasts could be related to more refined regional localities where configuration implies some degree of homogeneity of weather patterns. Such a reconstruction of 'weather regions' is fraught with difficulties and it is essential that areas retain adequate identity in the public eye.

Another persistent and major problem is the unevenness of the distribution of meteorological stations in Britain (Fig. 1.2). There is gross over-representation of low elevations and gross under-representations of medium and, especially, high elevations above sea-level. Of the major synoptic stations in recent use 40 per cent are below 100 ft (30·5 m) OD and 70 per cent are below 400 ft (121·9 m) OD. The density of the network of major synoptic stations decreases northwards and westwards from south-east England. This imbalance is all the more unsatisfactory when it is realised that more weather arrives from westerly points than easterly, on average. The punctual monitoring of changes in weather systems as they reach our coasts could be improved by the establishment of additional stations along the northern, western, and southern littorals of the British Isles. Errors in the predicted *timing* of weather events over twelve and twenty-four hours could thereby be reduced.

Availability of Personal Weather Forecasts

Personal weather forecasts are obtainable on request either to the Meteorological Office or, if available and required, private consultants. The latter are few in number in Britain by some international standards. A private forecast has the special advantage of being available only to the client who consequently may have an advantage over competitors. Stringer (1970) cites the example of some Midlands farmers who are prepared to pay for private forecasts on the basis of this argument. The Meteorological Office services are available to members of the general public on request; *pro rata* fees are imposed to cover the costs of abstraction of information. Buchanan (this book, Chapter 9, pp 115–25) demonstrates how weather inquiries have more than doubled in the nine years from 1962 to 1970 (Tables 9.1 and 9.2).

Fig. 1.2   Distributions of meteorological stations in the British Isles as at January 1969

(A) Meteorological Office observatories and synoptic stations manned by MO staff

(B) Auxiliary synoptic stations

(C) Climatological stations

(D) Anemograph stations

(E) Agro-meteorological stations

(F) Sunshine/rain and health resort stations

NB—Some stations combine several of the above functions
(Published with the permission of the Meteorological Office)

A

MAJOR SYNOPTIC STATIONS

KILOMETRES

MILES

AUXILIARY SYNOPTIC STATIONS

KILOMETRES
0 40 80 120 160
0 20 40 60 80 100
MILES

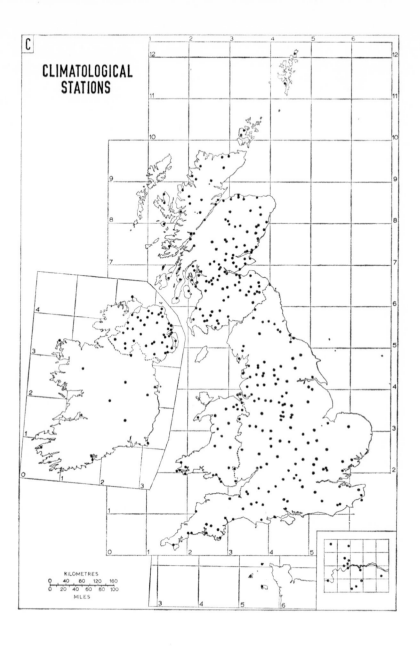

C

CLIMATOLOGICAL
STATIONS

KILOMETRES
0   40   80   120   160
0   20   40   60   80   100
MILES

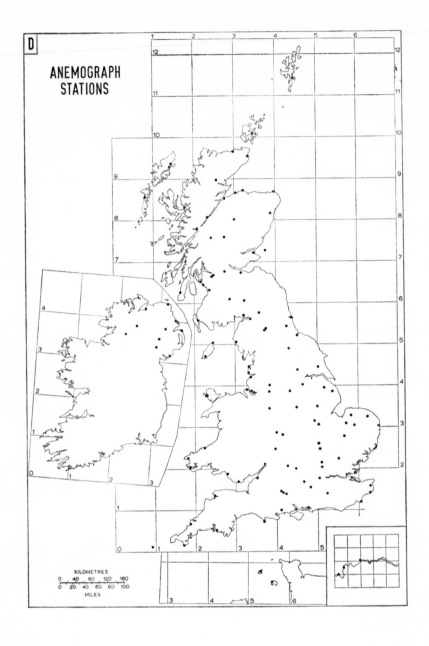

ANEMOGRAPH STATIONS

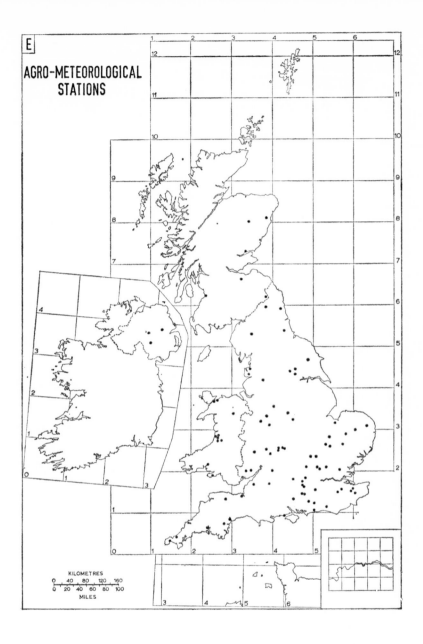

AGRO-METEOROLOGICAL
STATIONS

KILOMETRES
0   40   80   120   160
0 20 40 60 80 100
MILES

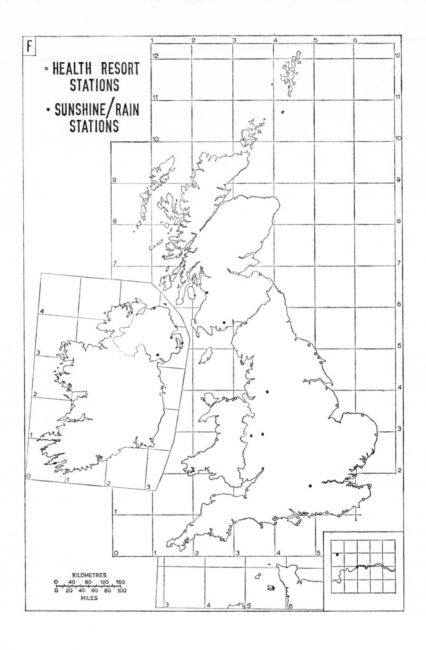

F

○ HEALTH RESORT
STATIONS

• SUNSHINE/RAIN
STATIONS

KILOMETRES
0   40   80   120   160
0 20 40 60 80 100
MILES

Here is concrete evidence of an expanding market for weather advice. The Meteorological Office has recently published (1970) a leaflet giving details of its weather services for the community (see Appendix, pp 232–42). If expansion continues at this pace, either the Meteorological Office section concerned will not be able to cope or it will have to be sufficiently augmented. On the other hand the work could be taken up by private consultancies, should they too increase in numbers and range, or particular industrial concerns may appoint their own full-time internal consultants. It would seem wise to recommend and anticipate an expansion of weather advisory services on all three fronts, viz, (a) within the Meteorological Office itself, (b) on a private commercial basis, and (c) as special appointments to particular companies. Whilst (a) would continue to underpin the service within the research environment of the Meteorological Office, (b) would not only provide healthy competition but also lead to a better appreciation of the services provided by (a) to the general public, and (c) would permit the necessary development of specialisation in the study from the inside of the weather sensitivity of particular industrial activities.

The Communication Media
Whilst the authorities of the press, radio, and television may well regard the time and space given to weather forecasts as adequate, it is in reality by several international comparisons, and indeed by its own standards, minimal. The point is supremely made by the forty-five-second glimpse of the television weather map just before 6.00 pm on weekday evenings or by the two-word weather summary squeezed into the newspaper stop-press column. Obviously, the communication media are in a powerful position in the flow of weather information between producer and consumer. They can expedite the flow or throttle it, and currently there is more evidence of the latter than the former. The television weather map now offers the most effective means of transmitting up-to-date and detailed weather information to the public. The style of presentation is showing steady improvement, especially on those programmes timed at 1.50 pm, including the forecast for farmers and growers given on Sundays at about the same hour. But these are hardly peak viewing times. Hogg (Chapter 6, pp 69–85) and Thomas (Chapter 7, pp 86–98), in this book, discuss the extent to which forecasts are used by farmers and suggest

an extension of the service and a re-timing to suit the convenience of farmers rather than the British Broadcasting Corporation. Without doubt the communication authorities are guilty of underestimating the demand for weather advice and it is also predictable that the allocation of more time and space to forecasts would lead to their more intensive and intelligent use.

Two-way Access between Weather Consultant and Individual Consumer
It should be stressed that *two-way* communication along the flows depicted in Fig. 1.1 is the most effective way of improving the system. It is essentially a matter of supply and demand. Already, horticulturists and other specialist growers have realised the financial benefits accruing from investing in a personal forecast from a consultant. Such problems as adequate notice for preparations to avoid frost-damage, knowing when to irrigate or spray against pests or diseases, or knowing what policy to adopt in variable weather at hay harvest, are more easily tackled if the appropriate local weather advice is available. In this sense, a weather forecast may realise its full value both commercially and functionally not least because inhibitive and monopolistic communication media have been bypassed.

Introduction of Weather-proofing
Weather-proofing is of three types, viz, (1) the use of physical devices for irrigation, shelter, etc, (2) the use of monetary cover via insurance policies, (3) the adoption of negative (eg, avoid the frost-hollow) or positive (sow hazard-tolerant varieties) strategies within management programmes. Duckham (1967) has shown how the productivity of grass-dairy farms in the English Midlands will vary not only with wet and dry summers but also with the implementation of fertiliser and irrigation treatments. The more intensive the system the more liable it is to oscillation in production in selected extreme years. The less intensive the system the less liable it is to such oscillations but over ten years, for example, continual underproduction must accumulate lower net profits. Frisby (1962 and 1963) has demonstrated the role of weather insurance against crop damage due to hailstorms in the Upper Great Plains of North America. It is cheaper and simpler to adopt monetary protection than to attempt to use physical devices which are expensive and difficult to implement effectively.

In western Pembrokeshire in Wales, the increasing competition in the early potato market has persuaded some growers to lay on irrigation which is applied during dry spells in the growing season (or early on in the season to avoid frost if necessary) to promote a continuously high rate of growth and to produce heavier yields earlier. It is not the *first* (earliest) man at the market who makes the largest profits; it is the first man with a substantially heavy crop who is the real pace-maker. Such a man is W. L. Thomas who writes (Ch. 8, pp 86–98) herein explaining that managerial finesse can dispense with the need to buy expensive irrigation equipment or thermostatically controlled chitting houses, and can at the same time side-step the weather hazards of early or late spring, frost frequency, etc. His system must be unique since he creates his own seasonal and even quinquennial weather forecasts but relies also on the general flow of public daily weather forecasts. By the controlled timing of the lifting of his own specially selected and home-grown seed potatoes, he can cater (a) for late or early development in the chitting trays during the winter, and (b) later on for early or late planting (in an early or late spring, respectively) in February. These decisions are based on his own prediction of winter and spring conditions. The recent run of mild winters and cold springs in Pembrokeshire was accurately forecast, and the carefully delayed programme prevented over-development of the seed potatoes in the chitting houses and achieved greater harmony with the lateness of the spring season. Profits were inevitably high especially when compared with those of other growers who planted over-developed seed too early. But Mr Thomas presents a unique case—one which makes many boxes in Fig. 1.1 quite redundant.

Education of the Consumer
The entrepreneur's own awareness of the value of using weather advice is clearly crucial. As the Director of the Meteorological Office has concluded (Mason, 1966), British weather forecasts are under-used and under-valued. This could be due to a persistence of the 'amateur' tradition and of the viewpoint that weather forecasts are more unreliable than reliable, and that little can be done about the weather. In fact, a great deal can be done, and current pressures to increase productivity are exposing the scale of weather costs especially in industries which operate outdoors (eg, the construction in-

dustries, agriculture, tourism, etc) or which use weather-sensitive techniques (eg, the manufacture of artificial fibres, the bakery industry, etc) or make weather-sensitive products (eg, foodstuffs and beverages). However, all industries using transportation are weather sensitive thereby and weather as related to health affects production via the labour force. The data provided by Buchanan (Tables 9.1 and 9.2) herein reveal the doubling of weather inquiries from industry between 1962 and 1970. Nothing succeeds like success and innovation may spread more rapidly by imitation than by any other means. The latter certainly applies to agricultural innovation. However, the real driving force in increasing consumer-demand is economic, eg, cost/benefit arguments which will precipitate a revaluation of weather forecasts and advice for both industry and agriculture.

Feedback in Weather Experience
A particular weather experience may condition a subsequent reaction to either the same weather situation or to some environmental factor common to the initial experience. Extremely good or bad weather during one particular holiday in one particular locality may affect a subsequent decision on whether to return for a second holiday or look for alternatives. Extrapolate this argument and a major factor in decision-making in tourism emerges which when evaluated on a 'resort', regional, or national scale could involve large sums of money. The guaranteed, hot, sunny weather of the Mediterranean together with economies of scale are the backbone of the package holiday in that area. A delay or deterioration in the summer season in the western Mediterranean area as compared with the central and eastern Mediterranean could *eventually* affect the comparative popularity of particular resorts. There is some initial evidence for such lateness and deterioration in the meteorological data for some western Mediterranean stations. It is possibly part of the current trend in seasonal climates, viz, colder winters, later springs, indifferent later summers, already identified by Lamb (1972) for the British Isles region.

Stringer's (1970) classical case of the effects of successive snowfalls on communications and commuting in the city of Birmingham, England, in December 1967 and January 1968 will bear repetition here. It shows in an amusing but also distressing fashion how the entire impact of weather is at the mercy of (a) *inter*-communication between public services and the general public as well as (b) the accuracy and communication of national and local weather forecasts. It also demonstrates admirably the need for a closer station network in northern and western parts of Britain (see p 6 and Fig. 1.2). Three separate snowstorms are involved, the first on 7 December 1967, the second on 9 January, and the third on 5 February, 1968. For the first, local authorities in the Midlands were warned by the local weather observatory to expect a snowfall of up to 3 in (7·5 mm) in the next twenty-four hours. In fact between 2.00 am and 6.30 this happened. Salting was started early but between 4.30 am and 5.00 am the snowfall was accompanied by a rapid fall in temperature. The first vehicles of the rush hour had only a shallow snow cover to deal with but it was frozen and they served only to pack it harder and create an ideal skid-surface. Gritting gangs were held up in traffic jams created by the immobilisation of vehicles especially at unsalted gradients adjacent to major road intersections. A total of 5 million man-hours was lost due to lateness or non-arrival of workers.

The memory of this chaotic morning stuck in the minds of commuters when the second very heavy snowfall occurred a month later on 9 January 1968. The local observatory predicted snow but the forecast put out by the Meteorological Office predicted a thaw. British Rail, following the national forecast, which proved incorrect, were unprepared for the snowdrifts of up to 1·3 metres which buried propane heaters and froze points, and suspended services between 7.00 and 9.00 am. Birmingham Corporation, however, relied on the correct local forecast, sent out salting and gritting lorries early and the roads were available for the morning rush-hour. Unfortunately, most commuters on seeing the heavy snow decided to leave their cars in the garage and travel by train! They flocked to trains which were not running whilst the roads had been cleared for a traffic rush which did not come.

The third snowfall, accurately predicted both locally and nationally, came on 5 February 1968, in the afternoon. Many commuters, mindful of the chaos of the two previous falls, went home early from work, but this very decision caused further chaos. The gritting lorries went out early but were caught in the 'early' rush hour. The railways were fully operational, ran to time, but carried only their usual 30 per cent of the commuters. What better illustration could there be of the need for full intercommunication between all parties concerned?

The Revaluation of Weather Forecasts

There is every reason to predict that the flow of weather information symbolised in Fig. 1.1 will intensify and become effective in reaching the consumer in the right form and on time. Already our weather forecasts are undercapitalised but increasing economic and social demand for their use will not only take up the slack but will also stimulate an improvement in the accessibility, accuracy, and applicability of the forecast statements. Thus advances in applied meteorology may eventually match those in pure meteorology to their mutual benefit, both in scientific and social terms.

References

DUCKHAM, A. N. (1967). Weather and farm management decisions, in Taylor, J. A. (Ed), *Weather and agriculture*. Pergamon Press, Oxford, 69–80.

FRISBY, E. M. (1962). The relationship of ground hail damage to features of the synoptic map in the Upper Great Plains in the United States. *J. Appl. Met.*, **1**, No 3, 348–52.

FRISBY, E. M. (1963). Hail-storms of the Upper Great Plains of the United States. *J. Appl. Met.*, **2**, No 6, 759–66.

GORDON, I. E. and BESTWICK, N. (1969). Understanding weather forecasts. *New Society*, **14**, 898–9.

LAMB, H. H. (1972). *Climate: present, past and future* I (*Part I: Fundamentals*). Methuen, London.

MASON, B. J. (1966). The role of meteorology in the national economy. *Weather* **XXI**, No 11, 383–93.

MAUNDER, W. J. (1971). *The value of weather*. Methuen, London.

METEOROLOGICAL OFFICE (1970). *Weather advice to the community*, Meteorological Office leaflet No 1.

STRINGER, E. T. (1970). The use of cost-benefit studies in the interpretation of probability forecasts for agriculture and industry: an operational example, in Taylor, J. A. (Ed), *Weather economics*. Pergamon Press, Oxford, 83–91.

CHAPTER TWO   C. V. SMITH

The Organisation of
Meteorological Advice for Agriculture
and Industry

> There cannot be a greater mistake than that of looking superciliously upon
> practical applications of science. The life and soul of science is its practical
> applications.—Lord Kelvin 1883.

The Management of Resources

'Science' may be defined as an activity aimed at moving back the boundaries
of human knowledge. This interrogation of nature is supported not because
there is necessarily a specific practical end point in view, but because it is
considered desirable in itself. 'Technology' may be defined as research and
developmental efforts aimed at specific practical goals. It is supported to the
extent that its objectives are thought to be important by its sponsoring body.
It is essential to keep the distinction between science and technology clearly in
mind. The objectives are different, the end products are different, and the
people and organisations involved need to be managed in different ways.

The consensus of opinion in recent years would appear to be that whilst
British science has continued to demonstrate its strength, British technology
has proved far from adequate. Whilst one must recognise that scientists and
technologists (in the sense of our initial definitions) are not necessarily
interchangeable, nevertheless those capable of making significant progress
when they reach the boundaries of present knowledge are likely to be the
exception rather than the rule. It could therefore be argued that the blame
for the neglect of technology when compared with science should be placed
squarely on the shoulders of those who have partitioned our scientific and
technological efforts in the recent past. To reiterate, technology will be

19

supported only to the extent to which its specific goals are seen to be important to its sponsors.

The problem of the management of scientific resources is not peculiar to the United Kingdom. Lithwick (1970) in his analysis suggests, for example,

> that Canada [too] has failed to harness science to economic growth or to solving social problems; and because there is no science policy coupled with national growth, areas of economic activity crying out for technological breakthrough will remain in the backwater. Decisions about the allocation of resources are in the hands of scientists and until there is a coherent science policy, stress will be placed on the new frontiers of science, because the alternatives lie outside the ken of this elite.

In this last phrase it seems that one major restriction to the advance of technology is identified.

Many of the scientists currently in office at universities and similar centres of learning, are not automatically or even necessarily familiar with problems outside their own specialisations, but still represent the obvious source of independent advice open to Government, its committees, and its institutions. If such advisers now continue to see their end results as new concepts expressed in learned papers, and if the allocation of resources is guided primarily by those who see speculative research as the only worthwhile study, then there is no *a priori* reason to suppose that operational research will function effectively at the same time.

There is competition for limited resources, men and money. The problems proposed for study arrive through quite different channels of communication in the two cases and if priority is allocated to those identified by conceptual rather than real-life situations, then already the only immediate opportunity of contact with the potential, practical user of the study has been lost. Another and quite substantial difficulty here is that real problems seldom come in 'discipline-shaped *blocs*' and the multidisciplinary approach required for their solution has occasionally been pronounced an anathema.

Both science and technology require individuals able to work at problems in depth within their own specialisation. The organisation of technology calls for something more. It requires a breadth of vision which extends beyond the individual discipline if the barriers and antipathies between the isolates of government and the financial community, industry, and university are to be broken down. The problems may have changed since 1883, when Lord Kelvin wrote the remark quoted at the head of this chapter, but the situation which prompted his remark is apparently still with us.

To emphasise the extent to which inadequate communication restricts the harnessing of science, we do not even need to stay with the natural sciences for further illustration. The social scientists of the United Kingdom have been scandalised to find that the simple publication of an academic analysis is no guarantee that its findings and implications will be acted on in the day-to-day world. The Social Science Research Council Report (1969/70) states that:

> It may be that the analogy drawn between the natural and the social sciences has led the practitioners of the latter to underestimate the extent to which effective research requires a much more active participation in the process of the diffusion of the results.

Evidently, the social scientists are discovering the hard way that even if the problems tackled are those to which others would like to have answers, then even when a satisfactory model of the situation has been set up, and simulated, there is no progress in the real world until results have been communicated back to it, in terms that it can understand.

The key to the successful harnessing of science, and the management of scientific resources to this end, would appear to be sufficient involvement in the problems of others in order to identify short-term objectives—involvement in the longer-term efforts needed to translate ideas, analyses, and understanding into effective action and change.

Objectives in Meteorology
Why and how do meteorologists develop an interest in agriculture and industry? The problems implicit here can be considered under three general headings. The first concerns the interrogation of nature—the exploration of those aspects of climate, weather, and environmental physics that are relevant. The second concerns objectives in meteorology. The third concerns the evaluation of environmental data and understanding—the translation of meteorological events into events of significance in other fields; a quantitative study of the interactions between the environment and various facets of production, storage, and distribution should lead on to rational and routine procedures to put this knowledge into action. These headings have been

placed in this sequence deliberately. The first presumably is not controversial; we have to start from the basis of facts, but unless our objectives comprehend the third type of problem, we do not even begin to become involved in the efforts needed to translate ideas, analyses, and understanding into effective action. The answer to the question 'Why an interest in agriculture and industry?' is simply that we place the user and potential user of meteorological advice at the centre of our deliberations; real problems are seldom strictly meteorological but we follow where they lead us.

In meteorology, both science and technology are presumably open to us. These activities are not mutually exclusive, but if we recognise a distinction between them, then we have to decide where our greatest effort should be placed. It is the personal view of the present writer that the following should be included among the general statement of aims of any meteorological organisation:

(a) incorporation of business objectives into meteorology;

(b) development of meteorology as a management aid.

Science has to be supported by society; it is not unreasonable for society to ask in turn that whatever dividends emerge of a practical nature should be extracted and exploited. Only the acceptance of some such objectives as the above will provide the working environment that enables this to be done.

Meteorology as a Business

If a business is a system for converting a resource, in this instance knowledge, into a contribution of economic value, then weather forecasting is surely a business. But a weather forecast has no utility of itself. Thus we are brought face to face with the simple business truth that within any organisation there are no results, only costs; results depend on people outside the organisation. To those who believe that meteorology is not simply for the meteorologist but for the benefit of the community at large, it is very apparent that the economic value of meteorological advice is nil until somebody, somewhere, takes decisions based upon it. For those outside meteorology, weather information is not an end product, but simply a starting point. Our output is a management input in other fields, and we need to develop meteorology as a management aid in order to make our output effective.

To write business objectives into meteorology is an immediate declaration of an interest in economic results. Our business has two basic objectives:

(a) marketing;

(b) innovation.

What is there to market in a monopoly public service (as meteorology is, with few exceptions, in the United Kingdom)? The purpose of a business is to serve a customer and marketing is the matching of the service given by an organisation with the customer's requirements, within the financial constraints laid down. This applies whether we are marketing food, textiles, or meteorological advice.

Innovation, to be justified, should result in the more economic provision of goods and services. It embraces improvement in the product, in marketing techniques, in management and organisation methods. It commits us to technological advance, rather than to problem-solving as this is generally understood. It allocates small priority to problems seen as difficult questions requiring tactical decisions for the restoration of normality. It allocates great priority to problems seen as situations which require that the right questions should be asked. It is concerned with finding out what the situation is, and what needs to be changed; and it achieves this by giving great thought initially to overall aims, and by setting down a succinct statement of long-term objectives, which then provides criteria for decision-making. Innovation is concerned with improvements in efficiency in any field. It could for example be logically argued that management techniques, developed for use in business, should have an obvious comparable value in the management of any large organisation. That they were first developed for business rather than to meet the problems of other organisations does not invalidate their more general applicability, but is simply a manifestation of the urge for profit; there is money in efficiency. There should be some benefit, perhaps not always measurable in money terms, from running a city well, or a health service, or even a meteorological service. Profit can be replaced by value for money.

How far would these business objectives go to meet the disparity between knowledge and its effective utilisation, which is the burden of the analyses made by the writers quoted? There is no doubt that they would go some way to reversing the priorities that have been given to 'science' and 'technology' in the past. To opt for such a change would not be to dismiss any interest in academic studies. The supposition here is that those capable of science (in our terms) can and would select themselves when brought up against technological problems. Implicit in innovation is the ability to see things in the round, to synthesise, to generalise, to question—to question accepted practice and to question fundamentals. Implicit in marketing is that the technological problems tackled should be selected by reference to industry and society. The aim is to provide not only the answers that industry and society need now, but also the answers they will require in five, ten, or twenty years' time. Marketing also implies that products and services should be seen through to a satisfactory working conclusion.

Agricultural Meteorology
In the organisation of meteorological advice for agriculture, there is an immediate and obvious distinction between the information obtained in the conventional, routine general weather forecasts and the work carried out by the agricultural meteorologist, who is concerned with the evaluation of meteorological and environmental data of all kinds.

Weather Forecasts
As applied meteorologists, we are committed to practical ends and to the provision of an effective service, which in an agricultural context is concerned as much with the longer-term weather prospects as an aid to the forward planning of field work as with the immediate weather of the next 12–24 hours. Now the need for quantitative statements does not diminish as the period of the forecast is extended; meaningful parameters such as rain-days or rainfall amounts are needed, rather than some indication of the future location of features on a synoptic weather map. One has to realise also that given meaningful meteorological parameters, the relationships between these and the possibility or advisability of field work may at present be known only to the agricultural meteorologist or to the agricultural advisory service.

Seasonal forecasts are not yet available. Monthly forecasts are made and published, but the statements they contain are fairly general. Weather is indicated by the departures from the climatological average of monthly mean rainfall and air temperature, but some attempt is also made to indicate the timing and sequence of changes of weather type that will inevitably occur over a period of about thirty days. The conventional monthly weather forecasts may represent the best guidance that can be offered. However, unless they can be made more quantitative than they are at present, it is unlikely that we can begin to be specific about weather-dependent activities such as field work. This suggests that we must content ourselves with the inferences to be drawn from the weather forecasts for a few days ahead, at most a week.

If our aim is taken as the interpretation of the longer-term weather prospects for the forward planning of field work, then the weather forecast for the coming week would appear to offer the most realistic scope for development. In addition we know that the past weather has a bearing on field operations both in the short- and long-term. The rainfall of last week may still dictate trafficability this week. The time to harvest will be determined in no small measure by the weather of the season so far. Weekly intervals would appear to be suitable and convenient periods for presenting summaries of past weather and commenting on the occurrence of critical or threshold values of meteorological parameters of some biological import on the farm.

The following schedule is put forward as a feasible undertaking to be included in a once-weekly broadcast aimed specifically at farmers (a break-down of the headings and possible topics for comment is shown in an appendix).

1 THE WEATHER SUMMARIES
   (a) The past week.
   (b) The season so far.
   (c) Comment on the success of the forecast and agricultural inferences over the past week.

2 THE WEATHER FORECASTS
   (a) The conventional weather forecast for the coming week.
   (b) The conventional weather forecast for the coming month.

3 THE AGRICULTURAL IMPLICATIONS OF THE FORECAST
   (a) For the coming week.
   (b) For the current season.

This project may appear over-ambitious but as applied meteorologists we have to follow where our problems lead us. It is only fair to point out that this schedule and the details of the appendix were developed in consultation with the Agricultural Development and Advisory Service (ADAS) of the United Kingdom.

For this service to develop satisfactorily, it is suggested that it must be approached on a regional basis. This in itself presents an initial problem since the British Broadcasting Corporation regions tend to divide the country from north to south, whilst a more natural division of farming interest might be from east to west. The requirements of the arable farmer are not necessarily those of the grassland or hill farmer or even of the horticulturist. In the early stages it would perhaps be essential to concentrate on an area of reasonably homogeneous farming activities so that user interests should not be dispersed by general statements (such as will inevitably occur if one attempts to cover the range of farming throughout the British Isles in a single broadcast).

Reality has to be brought into the programme. A reference to the success of the forecast of the previous week helps. A mention of the work actually carried out by an Experimental Husbandry Farm (EHF) might carry more weight. A reference to the projected work of an EHF over the coming week might serve as a useful reminder of the projects farmers should be currently prepared to tackle. Whilst on the subject of reality, a trial period (perhaps over a complete farming year) would be needed before one could consider presenting the information to the general public. Such a trial, and access to the EHF records, would obviously involve the ADAS of the region and entail the acceptance, by the stations and the individuals nominated to liaise with the meteorologists, of the aims and utility of the exercise. Such a trial period together with comment from the ADAS would also serve to identify where the present weather/field work relationships are inadequate.

Marketing Meteorology
The concept of the business approach is certainly not new to the agricultural meteorologist even if it has not been explicitly stated before; we have only to cite the early association in the United Kingdom of meteorologists with the agricultural extension service. Business objectives would argue at the very least for the maintenance, if not for the positive reinforcement of this present co-operation with ADAS. A business approach emphasises innovation rather

than problem solving; but assuming that we are to remain in close contact with the agricultural extension service, an organisation partly concerned with 'trouble shooting', it is inevitable that many practical problems presented to ADAS will extend the requirement for meteorological data and for the evaluation of the environmental physics of the situation. Seen in this light, problem solving is simply part and parcel of the marketing of meteorology. If, however, we are not to be managed by our 'in-tray' and the problems that are presented from day to day, not only must we have an overall set of objectives, but also a sensible staff structure to enable both the short-term (more mundane, but certainly economic) problems of the individual grower to be tackled, and also the longer-term strategic problems, which may have some repercussion on the industry as a whole. It is suggested that marketing, in the sense of problem solving, largely handles itself, provided that there are easy channels for physical communication between members of the extension and meteorological services and (and this is important) the people concerned are of the same calibre. Market research follows as a simple corollary.

The Management of Innovation
The continuing and central problem of management for agricultural meteorology is that of the management of innovation. In agricultural meteorology we are concerned with converting ideas into commercial reality (a management aid). We have the advantage over manufacturing industry that we do not generally have to produce hardware, and basically we are concerned only with ideas, their validation and marketing. (Our technical competence in research and development is assured if the staff is right.)

Now it is possible to find, in the literature, recommendations on the environment and organisation which are favourable to the generation of innovation, *vide* Casimir and Gradstein (1966). One can find check lists for creative thinking (Eiloart, 1969). One may examine case histories of particular innovations in industry (Cahn, 1970), and follow through the economic, political, and managerial arguments as to why particular projects (and countries) fail to do as well as others. One may do all these things and yet if one relies on serendipity then innovation may still fail. One way of improving this situation would be to institutionalise innovation, by deliberately making it part of the management function. Having set down the overall objectives, management has to formulate a summary of the strategies which will be followed to realise the objectives. These lead on to the tactical action required in the field of research and development, and marketing. At the tactical level we can now identify and emphasise the invention and innovation required to ensure the success of the strategies. Some method of sorting or ranking a variety of possible programmes is necessary in order to find which ones fit best with the available (manpower) resources.

Innovation follows on from 'science', but does not have to wait upon 'new science'. One useful strategy, possibly, would be the systematic exploration of our present state of knowledge for innovative ideas, bearing in mind current trends, likely developments, and the present subjects of research both in meteorology and agriculture, and also the type of problems that reach us. It is probably too much to expect that developments in one field will parallel the requirements in another, but we could use such an analysis to identify changes in the one that called for a complementary development in the other.

Let us take a few examples from meteorology. We have the advent of computers, the possibility of extended quantitative forecasts, the appearance of new observational techniques. Computers suggest that we should examine our procedures for data storage (and retrieval) and computation. Is our hand-processing of data still necessary? Has there been collaboration with outside interested bodies on the automatic 'reading' of script and figures? If our literature retrieval is to be based on a 'keyword' system, then does the World Meteorological Organisation have a project to produce an internationally agreed set of words? In some of our work we are concerned with the matching of meteorological and farm/biological data and a search for coincidences. How could this most conveniently be done by machine instead of as at present by hand? People are constructing 'models' for plant and disease growth, writing in day-to-day weather as a factor. What are the implications for our climate and agricultural industry? With computers available to do the calculation, what have we to offer to those interested in buildings and the weather-proofing of farm enterprises and farm products in general?

It may be a reasonable inference that the average farmer is capable of interpreting the conventional weather forecast for the next 12–24 hours in the light of his intended field work, the state of his crops, and the state of his ground. It may be no more than an act of faith that the average farmer can make use of conventional extended weather forecasts. Are we developing the necessary relationships between the meteorological parameters used (or likely to be used) in extended forecasts and parameters significant on the farm, ie, machinery work-days? New observational techniques are becoming available to monitor the atmosphere and its under-surface on a synoptic scale. How could we use satellite techniques, telemetry, and new sensors (for example, infra-red hand sensors) in our work?

In agriculture let us take simply one example, the work of the plant breeder. Selection is made on the basis of a few years' trial. Are we sure that the correct weather factors are being written into the data being compared? Were the years of the field trials of a particular variety extreme and unrepresentative of a longer-period sample of years? There are cereal varieties which do reasonably well in most seasons; there are other varieties which will do much better than 'reasonably well' given the right season. Can we influence the breeders' and the farmers' choice?

Conclusion

This symposium is primarily concerned with weather forecasting for agriculture and industry, but it is perhaps as well, at an early stage in the proceedings, to dispose of the misconception that weather forecasting is the whole of meteorology. The applied meteorologist is not concerned with the production of weather forecasts as such. His role is played before and after: before, in identifying the weather-sensitive problems that the grower and businessman must solve, and after, in communicating the implications of weather and environmental data to these people. To do this he needs to understand the interaction between weather (or climate or environment) and particular facets of industry. He needs validated working models of weather-sensitive situations in which 'weather' inputs lead to outputs directly applicable to managerial decisions in other fields. Weather forecasting now is not simply an exercise, the success of which is to be measured by the progress of lines across synoptic weather maps, but rather by the economic results of decisions taken under conditions of risk and uncertainty. To market meteorology successfully to business enterprises, the meteorologist needs to become involved, to the extent that business objectives are applied, quite naturally, to meteorology itself.

c

This paper tries to set down some of the attitudes and implications inherent in such an approach. In the light of the earlier quotations it emerges that these implications could, and perhaps should find more general acceptance in 'science-based organisations' as a whole. But it is only fair to say that if these attitudes in themselves would constitute an innovation for the organisation conceived as an entity rather than as the sum of its parts, then Haggerty (1970) would not be very sanguine about their successful implementation.

Acknowledgement

This paper is published with the permission of the Director-General of the Meteorological Office.

References

Anon (1970). Social Science Research Council *Annual Report*, 1969/70. HMSO.
CAHN, R. W. (1970). Case histories of innovations. *Nature, Lond.*, **225**, 693–5.
CASIMIR, H. B. G. and GRADSTEIN, S. (1966). *An anthology of Philips research.* Centrex.
ELIOART, T. (1969). Fanning the flame of innovation. *New Scientist*, 536–8.
HAGGERTY, P. E. (1970). Management of innovation. *Science Journal*, 5 April, 75–7.
LITHWICK, N. H. (1970). *Canada's science policy and the economy.* Methuen. (Reported in *Nature, Lond.*, **337**, 9–10).

Appendix

The Weather Summaries

(a) The past week

1 Rainfall amounts
Rain-days
Snowfall

2 Soil temperatures
Depth of frozen soil
Occurrence of freezing and thaw in the ground

3 Air temperatures
Number and severity of air and ground frosts

4 Sunshine hours
Potential transpiration soil moisture deficits
5 The occurrence of weather suitable for the spread of disease or infection, eg, Smith periods, Beaumont periods, Cereal Mildew infection days.

(b) The season so far

1 *Rainfall:*
    compare with normal expectation
    comment on soil moisture deficits (irrigation-need)
    comment on excess of rain after field capacity (leaching)
2 *Soil temperatures:*
    compare with normal expectation
    compare with normal expectation of first and last date of soil temperatures reaching 6°C
3 *Air temperatures:*
    compare with normal expectation
    compare with normal expectation of first and last frosts; average duration of frosts
    comment on accumulated temperatures (normal; time to harvest)
4 *Sunshine hours:*
    compare with normal expectation (time to harvest)
    potential transpiration
    effective transpiration (time to harvest)
5 The indications for the occurrence and severity of:
    (i) animal diseases
    (ii) insect pests and vectors
    (iii) plant diseases
6 Effect of season's weather on crop, in so far as this may be relevant to storage treatment, ie, time in store of top fruit

(c) The success of the forecast for
the previous week

Rain-days/work-days at EHF
(type of work)
Air or ground temperatures/
work-days of EHF
Changes of weather type

The Weather Forecasts

(a) For the coming week
A conventional presentation with synoptic maps.
Emphasis:
on changes of weather type
on sequence of rain-days
on the normal meteorological hazards in farming
eg, snowfall (exposure of stock, trafficability)
strong winds (exposure, physical damage, heating costs)
extended dry spells
extended wet spells
mild, humid spells

(b) For the coming month
A conventional presentation

The Agricultural Implications of the Forecasts

(a) The coming week
Given:
Rainfall amounts
Number of rain-days
Sequence of wet and dry days

We suggest:
the total number of days likely to be available for field work
the likelihood of being able to begin autumn ploughing
whether heavy land should be worked down or not immediately after
ploughing
whether spring ploughing, spring cultivations are going to be possible
(in conjunction with soil temperature)
whether the sequence of dry/wet days will favour germination or the
rooting of transplants (in conjunction with soil temperature, soil
moisture deficit, and possible wind-blow)

whether a spray programme will be possible or advantageous (in conjunction with soil moisture, sunshine, soil temperature)

whether it will be possible to harvest or carry field crops (in conjunction with soil moisture, soil temperature)

whether the carrying or application of fertiliser will be possible or worthwhile (in conjunction with soil moisture, soil temperature)

irrigation-need in general terms

Given:

Air temperatures
Soil temperatures   } absolutely or in terms of
Likelihood of air frosts   change from previous week

We suggest:

sowing dates

beginning and end of growing season

probable demand for frost protection devices for field crops

probable demand for heating of glasshouse crops (in conjunction with wind)

degree of exposure to stock kept outside (in conjunction with wind, rain)

possibility of conditioning grain or potato bulks

(b) The current season

Leaching—amounts of fertiliser to be applied

Time to harvest and advisability of taking on early harvest

Type of harvest—hay or silage—single or several fruit pickings

Amounts and quality of harvest—level of stockfeed and supplements for the coming winter

Possible or permitted run-down of stockfeed

Irrigation-need—whether in general terms irrigation is to be continued to avoid secondary growth

Time of return to field capacity—autumn work-days

Problems and Practice in Longer-Range Weather and Climate Forecasting

Climate is forever changing. The experience of one year differs from that of another. So does that of the decades and centuries. The variations from year to year may be illustrated by the mean temperature of the summer months (June, July and August taken together) in the lowlands of central England, which in about the last 100 years has ranged between 17·1°C in 1947 (17·6° in 1826) and 13·7° in 1879 (13·5° in 1860), figures normal for northern France (inland Brittany and Normandy) and Inverness-shire (Fort William), respectively. The mean temperature of the winter months has ranged from +6·8°C in 1868–9 to −0·4° in 1962–3, figures typical, respectively, for winters in south-west Ireland and southern Sweden. Averages for decades and longer periods are also liable to differences that are significant both in the proper statistical sense and in human affairs, agriculture, etc. The average temperatures in central England between 1920 and 1950 were 1·0°C higher in summer and 1·3° higher in winter than in the period 1680–1700, the time of the earliest thermometer records in Manley's (1959, 1961) carefully standardised, internally consistent series extending to the present-day. These figures are thought to correspond to an increase of about thirty days in the length of the growing season and a reduction in the average frequency of snow-covered ground from about twenty to five to seven days a year, as between 1680–1700 and 1920–50 in typical inland places in the lowlands of central England. Average temperatures in the 1960s were lower than in the early part of this century, 0·4°C below the 1921–50 values in summer and 0·8°C below in winter—with corresponding changes in the other items mentioned.

The variability of rainfall is of a magnitude no less important in a country where water resources are systematically exploited to the limit of their availability. The rainfall over England and Wales in the last 100 years appears to have varied from 139 per cent of the 1916–50 average in 1872 to 68 per cent in 1921. Over the area of Scotland the corresponding range is from 117 per cent in 1872 to 72 per cent in 1933. The 1916–50 overall average for England and Wales was itself apparently about 9 per cent higher than in the eighteenth century (1730–99) and 2 per cent above the last 100-year average (1851–1950); the 1916–50 average rainfall over the area of Scotland appears to have been as much as 8 per cent above the last 100-year average (1869–1968). The general upward trend of rainfall over England seems to continue more or less to the present time; though there are signs of a falling-off in northern and western parts of Scotland and Wales, and year-to-year variability has increased with the anomalous circulation patterns (and reduced frequency of westerly-type situations) in recent years. There was a run of dry years from 1961 to 1965 in England and Wales with only 89 per cent of the long-term average. In Scotland a run of dry years from 1885 to 1889 gave only 87 per cent of the 100-year mean.

These figures demonstrate the variability of Britain's climate on a variety of time scales and, hence, the need for forecasts to cover a month, a season, and (if possible) some years and decades ahead, to meet the demands of modern long-term planning in agriculture, industry, and international trade.

Scientific Approaches to Solution of the Forecast Problem

In the present state of knowledge, the standard methods of daily weather forecasting—extrapolation of the existing weather situation and computation of the forward development of the atmospheric circulation using dynamic and thermodynamic principles—cease to give a useful approximation to reality beyond about five days into the future. It is by no means certain that it ever will be possible to specify the situation by these methods beyond five to fifteen days ahead.

The difficulty arises as the prediction time increases in a number of ways: in principle, all are concerned with changes in the pattern and magnitude of the heating and cooling of the atmosphere, themselves due partly to changes of albedo (cloud, ice, and snow) and of ocean surface temperature, for which the atmospheric circulation directly is only partly responsible, and partly to external causes—which range from the regular, predictable seasonal changes of insolation to items such as volcanic and other dust and pollution in the atmosphere and fluctuations of the output of the sun. The external variables are presumably independent of the pre-existing atmospheric and terrestrial situation, though their impact may still be affected by it; the other variables are not so independent.

Forecasting methods other than those used in short-range weather fore-

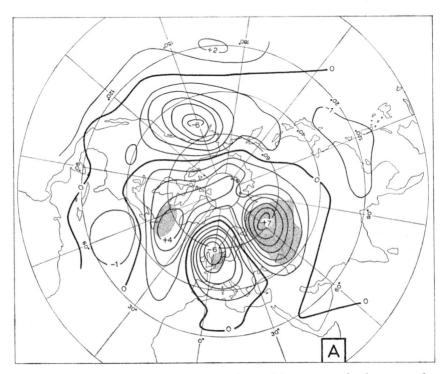

Fig. 3.1 Average anomaly in millibars of monthly mean sea-level pressure for February (*A*) following ten Januarys with warm sea, (*B*) *opposite page* following eight Januarys with cold sea (on average by 1 to 2°C) over the Newfoundland Banks area of the North Atlantic, 40–45° N, 40–65° W. Statistical significance beyond the 95 per cent level of probability is shown by stippling. (Data compiled by R. A. S. Ratcliffe and R. Murray)

casting therefore have to be employed and may always have to be employed. The approaches available may be listed as follows:

(a) through developing understanding of the *relationship between the physical, and especially thermal, condition of the earth's surface and the* large-scale characteristics of the *atmospheric circulation during the*

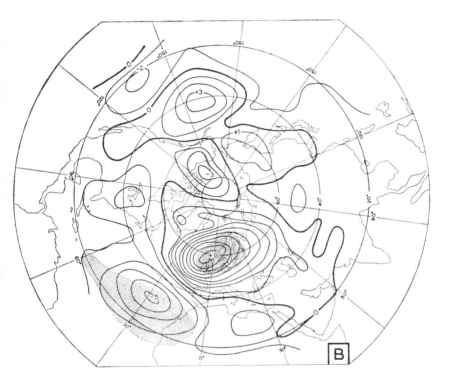

seasonal progression of the normal year—ie, of the general run of years—to understand how that relationship is operating, and what its effects should be, in the particular year or years with which we are concerned.

(b) study of the specific effects upon atmospheric circulation and weather of *anomalies of sea temperature and sea ice.*

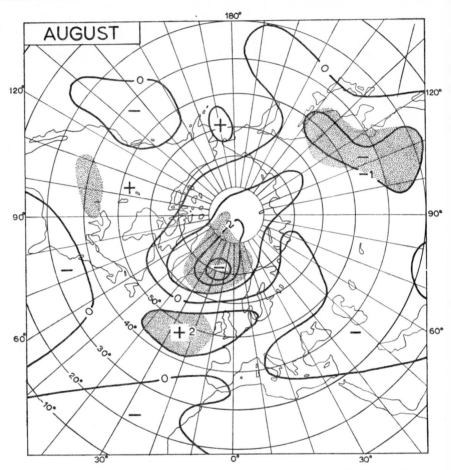

Fig. 3.2 Difference of monthly mean sea-level pressure in millibars in August, odd-numbered years minus even-numbered years between 1899 and 1964 (omitting 1939–46)—quasi-biennial oscillation. Statistical significance beyond the 95 per cent level of probability is shown by stippling. (Data compiled by R. Murray and B. J. Moffitt)

(c) study of the effects of anomalies in the *extent and distribution of snow.*

(d) study of any identifiable effects of the various kinds of *disturbances on the sun* and their course of development over a variety of time scales.

(e) identification of the effects of *volcanic and other dust and pollution* veils.

(f) identification of the time scales, manner of operation, and if possible the physical origins, of any *rhythmic or cyclic tendencies.* These appear to range in nature from tidal phenomena in the oceans, and apparent natural periods or life-cycles of some atmospheric (and, maybe, of oceanic) circulation processes, to variations in the tidal forces of the planets upon the sun and to items (or complexes of items) not yet at all understood. Cyclic tendencies around 30 days, 13–14 months, 2–3 years, $5\frac{1}{2}$, 11, 19, 22–23, 90, 100, 200, and 400 years (as well as some much longer cycles) may be accepted as playing an important part in the variations we observe—though most are only quasi-periodicities of somewhat variable period and amplitude.

(g) statistical *persistence tendencies,* perhaps mainly related to the large specific heat of water and the large amounts of latent heat involved in the phase-changes of water.

(h) statistically established *sequential tendencies* ('probable successions').

(i) identification of *analogous situations* in the past and monitoring how far parallel (or similar) development follows in the present and in past cases.

Use may be made under headings (a), (g), and (h)—and possibly other headings—of zonal, meridional, and other circulation indices and classifications of daily weather (and circulation pattern) type frequencies.

Working Methods
Confidence in forecasting depends upon identification of the physical processes at work, and (if possible) their interactions, and verifying this both by the dependability of the results in a sufficient number of past cases and by monitoring the course of the development in the current case. Analogous cases and statistical tendencies are only a reliable guide when founded on this identification. With or without physical understanding a statistical estimate of

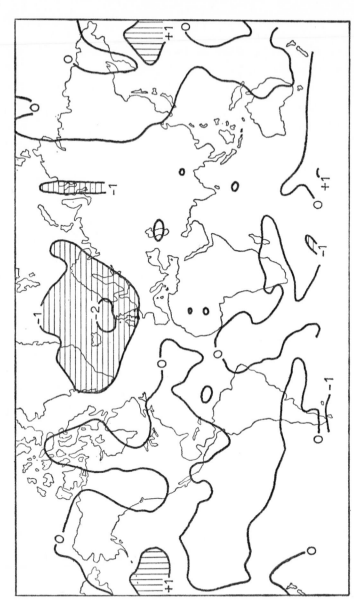

Fig. 3.3 Average anomaly of monthly mean sea-level pressure anomaly in millibars in Januarys two, three, and four years after sunspot maximum and two years before sunspot minimum, 1750–1958. (Data compiled by B. N. Parker)

the probability of any forecast is an aid to its application in practical decision-making, but lack of physical understanding may mean that some of the supposed analogies are false and the probability of this or that sequel is wrongly assessed.

Convenience is served by arranging forecast data in maps, prediction graphs, and tabular form. Effort is being increasingly turned in the Meteorological Office long-range forecasting research branch to casting the forecasting aids in these forms. Figures 3.1, 3.2, and 3.3 illustrate forecast maps which show the monthly mean sea-level pressure anomaly, and its statistical significance, given certain conditions of sea-surface temperature, the quasi-biennial (two- to three-year) oscillation in the earth's atmosphere and of the eleven-year sunspot cycle. Similarly, useful maps may be compiled of the percentage frequency of pressure anomalies of positive or negative sign with the given precondition.

Tables 3.1 and 3.2 are examples of two types of contingency table used in forecasting the winter's weather in Britain, based on 100 years' data between about 1870 and 1970.

Chi-squared significance tests can be applied to such tables for each precondition. Additional preconditions can be stated in terms of indices of daily circulation pattern (or weather) type frequency (Murray, 1970) or zonal index, etc over earlier periods of the year found to be important as indicators, eg, performance of the early December and early January westerly singularities in Britain and central Europe (Baur, 1958, Lamb, 1972).

Analogous cases for monthly forecasting (and beyond) are now selected by a variety of computerised scoring systems (including correlation techniques) under which each past year's relevance is rated under separate headings such as similarity of daily weather-type sequence over the previous month or season; though the final selection is made by the judgement of a panel, taking account of physical controls believed to be working on the circulation. The success of the forecasts is also reviewed after the event by a numerical scoring system (see, for instance, Ratcliffe, 1970).

Procedures which can properly be employed to gain useful insight into the probable climatic development over some years ahead must make use of the considerations we have listed (and should, so far as our understanding permits, take account of them *all*, giving appropriate weighting to each) in a

Table 3.1

Rainfall over England and Wales in terciles (R_1 driest third, R_2 middle third, R_3 wettest third of the past years examined). ('Rules' developed by R. Murray)

Preconditions in terms of central England temperature quintiles (T_1–T_5) and England and Wales rainfall terciles	Winter rainfall that followed Numbers of cases		
	R_1	R_2	R_3
Previous spring T_1	8	8	2
Previous summer T_5	3	6	10
Previous summer R_1	7	10	15
Previous autumn R_3	4	15	11
Previous October T_1	8	6	3
Previous November T_5	12	4	3

Table 3.2

Temperature over central England in quintiles (T_1 coldest 20 per cent of winters, T_2 next coldest 20 per cent, T_3 middle 20 per cent, etc). (Data contributed by H. H. Lamb and R. Murray).

Preconditions	Winter temperature that followed Numbers of cases				
	T_1	T_2	T_3	T_4	T_5
Last 10 days of the previous December (one year before) average temperature below 2°C at London (Kew)	5	5	4	1	1
Ditto in the winter to be forecast	10	4	2	0	0
Autumn R_1	5	4	6	6	11
Autumn R_3	8	11	4	6	3
September and October in Scotland and England and Wales R_1	2	3	3	6	6
September and October in Scotland and England and Wales R_3	3	11	4	0	1

manner similar to the maps and tables here illustrated from monthly fore-casting practice, but also with extrapolations of quasi-periodicities and ultimately reviews of analogous cases, followed by monitoring of the de-velopment up to the time of issue of the forecast verdict. It is a defect of most climatic forecasts so far issued that they have each been based on analysis and extrapolation of a single influence (solar, tidal, or unexplained cycles). An example of a forecast based on rather wider argumentation (a proposed 200-year cycle in wind direction frequencies in NW Europe, tentatively identified with a cycle in solar and atmospheric radiocarbon data) is given in Weiss and Lamb (1970). The huge economic stakes involved in many de-cisions dependent upon assumed future climatic values make it essential that full data, and an outline of the physical argumentation, should be produced openly to the interests concerned in a form that facilitates assessment of the reliance to be placed upon any forecast judgement and the probabilities of alternative sequels. This manner and fullness of presentation is also the client's main guarantee of the relevance and soundness of the information proffered: for this is a field where ill-informed consultants and organisations may be tempted to offer their services at high fees.

References

BAUR, F. (1958). *Physikalisch-statistische Regeln als Grundlagen für Wetter- und Witterungsvorhersagen*, II, Akad. Verlag, Frankfurt/Main.

LAMB, H. H. (1972). *Climate: present, past and future*, I (*Part I Fundamentals*). Methuen, London.

MANLEY, G. (1959). Temperature trends in England, 1698–1957. *Archiv. f. Met. Geophys. Biokl.*, B9, 413–33.

MANLEY, G. (1961). A preliminary note on early meteorological observations in the London region, 1680–1706. *Met. Mag.*, 90, 303–10.

MURRAY, R. and BENWELL, P. R. (1970). PSCM indices in synoptic climatology and long-range forecasting. *Met. Mag.*, 99, 232–45.

RATCLIFFE, R. A. S. (1970). Meteorological Office long-range forecasts: six years of progress. *Met. Mag.*, 99, 125–30.

WEISS, I. and LAMB, H. H. (1970). Die Zunahme der Wellenhöhen in jüngster Zeit in den Operationsgebreten der Bundesmarine, ihre vermütlichen Ursachen und ihre voraussichtliche weitere Entwicklung. *Fachliche Mitteilungen*, Nr 160. Geophys. Beratungsdienst der Bundeswehr, Porz-Wahn.

CHAPTER 4   J. P. HUDSON

Agronomic Implications of Long-term Weather Forecasting

Unpredictable vagaries of crop yields pose one of the biggest problems in agriculture and horticulture, making it difficult to develop a rational system of marketing, and variations in the weather seem to account for a high proportion of the fluctuations in yield. Of course, some managerial decisions do not depend on the weather, but many do, and in the absence of adequate forecasts the *weather-dependent* decisions must necessarily be based on practical experience and on intuition. These can be inadequate (and sometimes expensive) guides to decision-making in the increasingly complex business of crop husbandry and various attempts are being made to rationalise the decision-making processes. It seems possible that longer-term weather forecasts, of predictable reliability, may become available in the not-too-distant future, and it is timely to consider what use (if any) could be made of such forecasts, if and when we get them.

A paper on this subject was given in a Colloquium on Crop Geography and Adaptation, held during the 17th International Horticultural Congress, Maryland, USA (Hudson, 1967), since when there have been important developments in both meteorology and crop physiology. The purpose of the present paper is to bring the review up to date in the light of these developments.

Causes of Variation in Yield
One of the characteristics of crop husbandry is that yields tend to vary widely, uncontrollably, and unpredictably from year to year. There are several sources of variation for any particular crop:

Variation	*Probable cause*
From field to field, on the same farm (ie, under the same management)	Soil and microclimate
From farm to farm, on the same soil and type of site	Managerial capacity of the farmer
From year to year, on similar fields (ie, same soil and site, and standard of management)	Weather

Fluctuations in yield tend to be largest in some of the crops that are most complicated and difficult to grow, such as fruit, yet the demand for such crops is relatively static, and heavy yields often produce gluts that wreck the price structure. There is, therefore, a strong incentive to even out yields in such crops, so that marketing is a more orderly process.

At the same time there is need to increase average levels of yields, without commensurate increases in cost of production. It is the aim, for instance, of the Agricultural Research Service, to provide the information needed to enable agricultural production in the United Kingdom to be doubled, from a diminishing acreage, by the end of the century.

Altering Weather, Plant Responses, and/or Management?

There are three possibilities, so far as the weather is concerned, namely (a) to alter the weather, (b) to change plant responses to the weather, and (c) to adjust management more skilfully to the weather. Regarding the first, there seems to be little likelihood of altering the macro-weather to suit farmers or growers in this country, despite some progress abroad in the dissipation of cold fog and the suppression of hail, and a good deal of inconclusive work on rain-making. On the other hand, of course, horticulturists already go far towards altering the micro-weather by the use of irrigation, shelter, shade, mulches, and similar devices.

The second possibility is much more interesting. There seems to be a real chance of stimulating plants to grow better under sub-optimal weather conditions by the use of exogenous growth regulators when we understand more about plant physiology (Hudson, 1958), and this is a field where there have been rapid developments in the last few years. For instance, before long it should be possible to alter the balance between shoots and roots, by the use of appropriate fertilisers and/or growth substances, to 'pre-condition' plants so that they will yield better under relatively adverse weather. This might be achieved by encouraging a deeper, better-branched root system if relatively dry weather is expected, so that soil moisture is better exploited, or even reducing roots relative to leaf surface where the evaporation rates are likely to be low.

However, the immediate interest lies in the third possibility, that we should adjust managerial practices more skilfully to reduce some of the fluctuations in yield that are at present caused by adverse weather. This will require a deeper insight into plant/weather relations than we have at present.

Forecasting—The Present State of the Art

It used to be thought that *climate* remained virtually constant over periods of centuries, and even thousands of years, but careful analysis of long-term records since the 1940s has shown that there have been significant changes, even in the past seventy years and especially in the 1960s, with first a general warming and now a cooling off (Lamb, 1969). The biological implications of such climatic changes must be complex, but immensely important to both global and local patterns of food production. Within the next thirty years it may be possible to predict long-term climatic changes from physical models (Mason, 1970), but that development is still a long way off.

At the other end of the scale, there has been a noticeable improvement in the quality and accuracy of short-term weather forecasts in the last few years, and further improvements seem to be imminent. For instance, quantitative, and quite detailed, forecasts of rainfall on a daily basis will probably be available in 1972. Five-day forecasts have been issued twice a week for a number of years in the USA, with an increasing standard of accuracy, and we shall probably have usefully reliable forecasts for a week ahead within the next few years.

Forecasts for a month ahead are continually being improved, and these forecasts are now rarely totally misleading. In the even longer-term, experimental seasonal forecasts are being made in the Meteorological Office by extending the techniques used for the monthly forecast, and 'the last seven forecasts have all been encouraging' (Mason, 1970).

In passing, it would be useful to have such information as the probable length of the frost-free period, or the probable sunniness of the summer, but such information would be of maximum use only if we also measured, and took more account of, current attributes of the soil such as (a) the amount of water available at the beginning of the season in the potential rooting zone, (b) the heat status of the soil, especially at 0·3 to 0·6 m, and (c) the nutrient status of the soil, especially if the previous season was wet enough to have led to significant leaching (van der Paauw, 1968).

Agronomic Uses of Weather Forecasts
With the possible advent of longer-term weather forecasts, of increasing reliability, it behoves agronomists to consider how these could be used in management, and to provide the information that will be needed in order to take full advantage of such forecasts when they come. The managerial implications of forecasts will depend very much on their period, ranging from a few days to a few years, and periods of five days, five weeks, five months, and five years have been chosen for consideration here, since these seem likely to be of special managerial significance. The following aspects are worth considering: they are mainly taken from the general field of horticulture, since other contributors are dealing with the broader agricultural field.

Five-day Forecasts
To determine the depth at which seed should be sown to achieve optimal rate of emergence and to ensure the precise spacing of a pre-determined plant population, essential for heavy yields in a 'weed-free environment' with no post-emergence cultivations.

To determine whether or not to sow a crop (eg, as closely argued by McQuigg —see pp 49–50).

To allow for changes of soil temperature where resistance of young seedlings to fungal attack is temperature-sensitive.

To take account of rainfall expected in the next few days (especially when *amounts* can be forecast) when deciding whether or not to irrigate under marginal soil moisture conditions, especially where water is scarce.

To time the destruction of potato haulm so that there is maximum protection against blight and minimum loss of crop.

To decide whether or not to harvest some crops.

To ensure maximum efficiency of spray programmes, where incidence of disease and pest is affected by weather.

To prepare in time for the protection of crops against frost.

To increase the efficiency of some herbicides.

To decide whether or not to supplement natural pollination of fruit in poor setting weather.

Five-week Forecasts

To extend the decision regarding time of seeding to cover successional sowings of crops like peas, to ensure an even flow to the cannery or freezer.

To harvest crops for short-term storage where adverse conditions are likely to interfere with lifting (eg, carrots frozen in the soil) but where storage is not usually economical or necessary.

To decide whether to put perishable crops into short-term storage, to even out supplies to the market, based on estimates of demand and supply.

To decide on future use of water, where this is scarce, based on the probability of non-moisture-stress days (Denmead and Shaw, 1962).

To decide on dates of sowing crops sensitive to frost, to maximise the chance of achieving a large leaf-area index by the time light conditions are optimal (Watson, 1947).

To avoid spraying when diseases or pests are unlikely to be troublesome.

Five-month Forecasts

To decide whether or not to grow marginal crops—eg, outdoor tomatoes in England, where they are a complete loss in a cool summer but highly profitable in a hot season.

To aid in the management of scarce water resources.

To determine schedules and sizes of contracts for crops and labour, and the timing of cropping schedules.

To match fertiliser applications to expected yields.

To plan timely measures against diseases and pests likely to be favoured by the expected season.

To choose varieties most likely to thrive in the expected pattern of weather (eg, 'warm-summer varieties', 'wet-season varieties', and so on).

To determine the acreage needed to give a required tonnage of crop.

Five-year Forecasts

To decide on the general pattern of investment, especially in equipment to counter effects of adverse weather; eg, for irrigation, frost protection, short-term storage, and grain drying. Avoidance of overinvestment could be as important as ensuring adequate cover.

To determine national policy with regard to support, or otherwise, of de- delopments in marginal areas.

To determine the choice of crops to be grown, where there are alternatives favoured by different types of weather, eg, farmers might switch from cereals to grass or vice versa.

Increasing attention is being paid by research workers to some of these topics. For instance, Smith (1970) has discussed the value of including the expected sequence of wet- and dry-days within a seven-day forecast, since rain tends to interfere with many outdoor farming operations. One of the most interesting approaches in this field has been by McQuigg (1968) whose methods are based on a combination of observations of present conditions and forecasts of future weather. A good example is his schedule for sowing cotton, based on the following series of questions and answers:

Should we sow cotton *today* or not?

Is soil dry enough to get on the land?

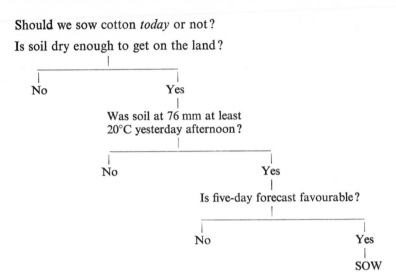

Hashemi and Decker (1969) concluded that data on weather in the past, coupled with forecasts of rainfall and evaporation rates in the future, could be used to programme irrigation more efficiently, with a significant saving in water during the summer when the need for supplemental irrigation is most acute.

In the longer-term, a UN meeting in Rome (reported in *World Hunger*, November 1970) suggested that long-range climatic trends should be taken into account when planning agricultural development projects (eg, if the global trend towards a warmer climate is believed to be slowing down, farmers should be actively dissuaded from planting citrus in areas where *near freezing* temperatures have often been recorded in the recent past, as there may be *frosts* in future years). Indeed, the success or failure of whole farming systems may be in the balance if the climate changes even marginally over a period of a few years, as many pioneer farmers have found in the past. In view of recent doubts cast on the validity of the thirty-year average (Lamb, 1969) as the basis for assessing agro-climates, the value of many surveys of the agricultural potential of semi-arid areas must be in some doubt.

Interactions between Varieties and Weather

In the past, farmers and growers have, of necessity, developed systems of production that are relatively insensitive to the ranges of weather that occur in their locality. This situation may well change. For instance, the choice of varieties to suit expected seasonal weather has an obvious economic implication. Such a choice would require (a) weather forecasts of predictable reliability, covering those parts of the season when weather is critical, and announced early enough to enable the choice of variety to be made before sowing, and (b) much greater knowledge of the responses of varieties to weather than is available at present.

In the absence of long-term forecasts growers have tended to be conservative in their choice of varieties, preferring to grow those that are 'weather-tolerant', and reasonably certain to give useful yields whatever the weather turns out to be, rather than to gamble on varieties that might do particularly well in one type of season but not in another (Hudson, 1965). This attitude might well change with the advent of reliable long-term forecasts, if it is known which seasons 'favour' different varieties.

There is remarkably little information about such effects, despite the vast amount of effort that has been spent on testing varieties, because the methods used at present are not designed to reveal variety × season interactions. Indeed, most of the methods used in variety trials are based on taking averages over a number of seasons, usually at least three, to damp out the effects of differences in seasonal weather, and much longer trials would be needed to reveal the effects of particular types of weather. However, where such data are available, some interesting interactions have been revealed. For instance, Bunting and Willey (1958) found that the yields of two varieties of maize were significantly reversed by summer weather, cool summers favouring one and hot seasons another (Fig. 4.1). In very large-scale trials in the Sudan (Siddiq, 1966), there were highly significant differences between two varieties of cotton, the yields of which were reversed in different years (Fig. 4.2).

The relation between plants and the weather is a complex matter, and it may take a number of years to acquire the information on which decisions could safely be based regarding which variety to sow for any particular type of season. It is therefore not too early to consider what should be done to

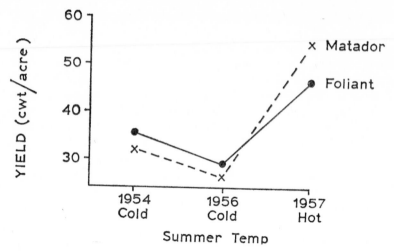

Fig. 4.1 Yields of two varieties of maize in three different seasons (after Bunting and Willey, 1958). Differences significant at $P = 0.05$. (Reproduced by kind permission of the NIAB)

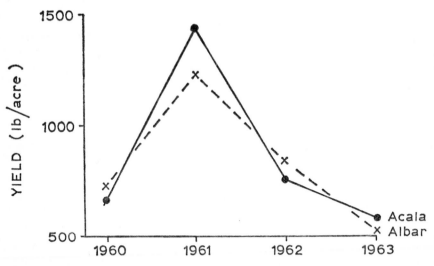

Fig. 4.2 Yields of two varieties of cotton in four different seasons in the Sudan (after Siddiq, 1966). Differences significant at $P = 0.01$

study varietal responses to weather, even though long-term forecasts may not be available for some years to come.

Plant breeders are, of course, fully aware of the importance of the effects of weather on yields and are constantly seeking to produce varieties that are better adapted to environmental conditions, including weather. There does not, however, seem to be much interest in producing ranges of varieties to match particular weather patterns. Indeed, in vegetables for processing, where an even flow to the factory is so important, there is increasing pressure to select varieties which are particularly *tolerant* of weather fluctuations. There seem to be no readily available data on the importance, or even the existence, of variety × weather interactions relative to other factors such as the nutritional status of the soil, yet these are obviously of great importance.

Conclusions

Agriculture is one of the most weather-sensitive of all industries, yet we are still nowhere near a fundamental understanding of the effects of weather on yields. Geographers often know more than agronomists about agricultural and horticultural cropping systems, and farmers know far more about methods of crop production than about the relations between crop plants and their environment. The present lack of understanding has led, unconsciously, to the development of general-purpose varieties and methods of production that are relatively insensitive to the weather, but these will not be good enough when more is known about the future weather.

McQuigg has suggested that improvements in the accuracy of weather forecasting will result in increased economic returns if (a) the decision-maker has alternative choices of action, in which there is a real difference in the economic consequences of at least two of them and (b) if he understands the relation between these events and his crops well enough to allow him to choose the most profitable alternative with certainty.

Five-day forecasts may sometimes be highly important, and sometimes not at all, in the sense that they may not lead to a change in plans. *Five-week* forecasts may prove to be of strangely limited value, but the writer's opinion is that growers could reap big benefits from *five-month* forecasts, provided they were available in time, before crops are sown. In the long run such forecasts might enable varieties to be sown that are likely to do particularly well in the sort of season that is expected, eg, a cool summer, a warm spring, a sunny August, and so on. However, in saying this one must hasten to add that we do not yet have the necessary information about varieties on which to base the choice, because the present methods of testing varieties are designed to buffer seasonal effects and identify those varieties that are likely to be least affected be adverse weather, rather than those favoured by a particular type of season.

The general conclusion is that a better knowledge of forthcoming weather might add considerably to farming efficiency. It is doubtful if we shall ever improve much on the yields now obtained by the best growers in the best years, but it should be possible to raise the general average and to reduce very much the incidence of gluts and shortages, thus facilitating more orderly marketing of the perishable types of produce.

Meteorologists are particularly vulnerable in their work on forecasting since they cannot hide their inevitable mistakes as well as the rest of us! They are therefore likely to be cautious in referring to longer-term forecasts, which may thus be presented, at the appropriate time, as a *fait accompli*. Agronomic research usually takes a long time to complete. If longer-term forecasts turn out to have agricultural implications there may be a long delay in making use of the new information if research is not started well in advance of the time when it is needed. It behoves us to consider the implications of long-term forecasts, and to start any necessary investigations *now*, to avoid finding that an important new factor has emerged in crop management which we cannot use because we have not prepared ourselves for it.

References

BUNTING, E. S. and WILLEY, L. A. (1958). Studies on maize for grain production. I. Variety trials. *J. Nat. Inst. Agric. Bot.*, **8**, 364–77.

DENMEAD, O. T. and SHAW, R. H. (1962). Availability of soil water to plants as affected by soil moisture content and meteorological conditions. *Agron. J.*, **54**, 385–90.

HASHEMI, F. and DECKER, W. (1969). Using climatic information and weather forecasts for decisions in economising irrigation water. *Agric. Met.*, **6**, 245–57.

HUDSON, J. P. (1958). Effects of weather on plant behaviour. *Nature, Lond.*, **182**, 1337–40.

HUDSON, J. P. (1965). Agronomic implications of long-term weather forecasting, in *The biological significance of climatic changes in Britain*, 129–34. Academic Press, London.

HUDSON, J. P. (1967). Horticultural implications of long-term weather forecasts, in *Proc. XVIIth International Horticultural Congress, Maryland*, 1966, 63–76.

LAMB, H. H. (1969). The new look of climatology. *Nature, Lond.*, **223**, 1209–15.

McQUIGG, J. D. (1968). A review of problems, progress and opportunities in the use of weather information in agricultural management, in *Agroclimatological methods*. Proceedings of UNESCO Symposium, Reading, 1966.

MASON, B. J. (1970). Future developments in meteorology: an outlook to the year 2000. *Quart. J. Roy. Met. Soc.*, **96**, 349–68.

SIDDIQ, MHD. A. (1966). Variation in performance of varieties of cotton in the Sudan Gezira. *The Empire Cotton Growing Review*, **43**, 98–106.

SMITH, C. V. (1970). Weather and machinery work-days, in *Weather economics* (J. A. Taylor, ed). Pergamon Press, Oxford.

VAN DER PAAUW, F. (1968). Aspects of soil fertility in weather-crop-yield relationships, in *Agroclimatological methods*. Proceedings of UNESCO Symposium, Reading, 1966.

WATSON, D. J. (1947). Comparative physiological studies on the growth of field crops. I. *Ann. Bot.*, **11**, 41–76.

CHAPTER FIVE

A. N. DUCKHAM

Meteorological Forecasting and Agricultural Management

The function of forecasting is to enable the decision-maker to take better informed positive action, evasive action, or no action at all. For this purpose, in agricultural management, projections into the future of correlations observed in the past between meteorological events and the apparent biological and/or economic consequences of such events, may be of great value. This is largely because such extrapolations, even if they can be expressed only as statistical probabilities, are in effect long-term forecasts. They are needed for strategic decisions. Short-term forecasts are only of tactical use. This paper is not, therefore, about tactical day-to-day decisions, ie, not about the value of weather forecasts in the use of resources already committed. It is about strategic agricultural management-decisions that are (or at least could be) designed to avoid, mitigate or exploit predictable or probable weather or weather-induced conditions. It concerns the use of weather or climatic data to make predictions that the decision-maker can do something about in reaching investment or long-term resource-use decisions.

In the present context, 'meteorological forecasting' includes biological or economic forecasts that are only partly based on recent weather data or on probabilities derived from longer-term past weather records; 'agricultural management' includes the management of firms that supply inputs to, and that buy and process outputs from farms, and governments and their agencies as well as farms and farmers.

Medium- and longer-term weather-based agricultural decisions fall into three classes:

(a) actions based on 'early warnings' of meteorological origin; these are mostly seasonal, ie, less than twelve months.

(b) land-use and enterprise selection; this is usually medium- or long-term, eg, two years and upwards.

(c) investment in 'weather-proofing' or other environmental improvements, eg, in irrigation or livestock housing: this is frequently truly long-term, ie, decades rather than years.

Early Warning Systems: the Forecasting of Biological Consequences

Any biological species must optimise its resource use so as to survive and, if possible, expand, eg, by beating competition. In many species, including man as a farmer, this optimisation depends *inter alia* on the species' ability to predict or identify adverse or favourable seasonal conditions including departures from seasonal norms. The biological principles involved in such processes are well illustrated by D. Koller, R. Levins, P. F. Wareing, C. M. Williams, and others in Woolhouse (1969). Thus, amongst deciduous woody species showing photo-periodic responses, bud dormancy is normally hastened by short winter days, whilst the breaking of dormancy depends in some species, eg, beech (*Fagus sylvatica*), on long days only. Other species respond exclusively to chilling or to a combination of chilling and longer days (Roberts, 1970). In this way, leafage is avoided in winter conditions.

Cool temperate breeds of sheep come on heat only as the result of shortening days so that, allowing for five months' pregnancy, lambing does not take place until the weather is milder and some grazing is likely to be available. Coat-shedding in cattle, ie, protection against cold, is influenced by day-length. Many other examples of such photo-periodic and other meteorological conditioning or 'early warning' systems could be quoted for both wild and economic species (Duckham, 1963, Chaps 1, and 8 to 11).

Farmers have, of course, for many centuries recognised the inevitability of, and taken steps to avoid or reduce, adverse seasonal meteorological conditions, eg, by the winter housing of livestock in cool, wet climates and by irrigation in hot, dry climates. More recently the control of seasonal photo-periodic effects, eg, on laying poultry, on the flowering and consequent loss of nutritive value in grasses, on some decorative flowers (cut-plants) has become common.

Most farmers and traders are familiar with early warning systems, eg, a very late spring usually foretells low barley yields, and a wet and delayed hay harvest reduces the digestibility and the protein and mineral contents of hay which has later adverse effects on winter milk yield unless the cows' winter diet is suitably boosted. But progress with 'early warning' notices has recently been rapid on less predictable weather-induced problems. In the United Kingdom, the work of L. P. Smith and his colleagues in the Meteorological Office in conjunction with plant pathologists or veterinary research workers has provided some notable 'early warning' devices. These mostly come, or could come, early enough for the farmer to take avoiding action. Even if they are too late for action, such predictions of risk could be useful to the farmer (eg, to warn him of lower sales receipts or to buy in more concentrates for winter feed), to the trader and food processor, to wholesale markets, and to the government.

More progress in prediction seems to have been made with animal diseases where the time-lag between meteorological event(s) and their pathological manifestation is (Smith, 1970) largely independent of weather after the critical event(s). Thus, *liver fluke* in sheep can, on the basis of summer rainfall and transpiration, be forecast in August for disease incidence three to six months later. In *Nematode* (round worms) infestation in lambs, the date of worm hatching and the severity of the disease in April/May can be foretold from March earth temperatures. *Gastro-enteritis* in cattle, which is partly weather dependent, can be indicated, by October, as a risk next May and June (Smith, 1970). Amongst nutritional effects the percentage of *twin lambs*, which tends to rise if the ewes are nutritionally flushed before being put to the ram, can be meteorologically predicted in hill-sheep country by autumn tupping time, whilst the risk of *pregnancy toxaemia* and *hypomagnesaemia* in ewes may also prove to be meteorologically predictable (Smith, 1970).

In crops, *virus yellows* (of which the vector is an aphid) in sugar beet can be forecast usefully (Smith, 1970) from weather data, and so of course can *potato blight*, but here the warning is short, almost in the tactical class.

At the broader commercial levels, the link between weather, nutrition, and metabolism makes it possible to predict milk yields in the UK for twelve months ahead using November–March milk output and March soil temperature values (Smith, 1968). Official crop yield forecasting, partly based on weather data, is commercially important in North America. In the Sudan Gezira, cotton yields are largely dependent on moisture conditions before planting; these can be recorded and used to predict the size of the crop (Wrigley, 1969, p 74).

The explanation of some of the above 'weather chain' effects is obscure. But neither the empirical nature of the predictive formulae nor the pragmatic foreknowledge of farmers or traders reduces the need for more joint research by meteorologists, agriculturalists, and applied biologists into possible weather chains so that meteorological 'cause' and biological 'effects' can be quantified and used with greater precision as early and public warnings.

Land-Use and Enterprise Selection

Meteorological feasibility studies in the USA, Australia, New Zealand, East Africa, and elsewhere 'have shown the value of probability data in deciding land-use and animal stocking programmes ...' (Duckham, 1966). At first sight it would seem that the use of predictions based on analyses of past weather records would be useful only in opening up new lands. Then answers have to be found to such questions as 'Is this land suitable for agriculture rather than forestry or is it best left as it is?' or 'If it is suitable, what enterprises (specific crops, grass), what farming system (eg, tillage or alternating), and what input intensity (eg of fertilisers) should be adopted?'

Prima facie, this approach would have no role in areas where, over decades or centuries, farmers have, by trial and error, found answers to these questions and evolved a settled but not necessarily stable form of agriculture. But, apart from the high probability that science and technology will supply new and revolutionary techniques (eg, the new 'miracle' rices and dwarf wheats) that require re-assessment of existing land-use patterns, there are new pressures of social, economic, political, or strategic forces, such as the explosive growth of population in tropical developing countries and the increasing demands for land for recreation, housing, etc in densely populated industrial countries. Thus, when the UK enters into the European Economic Community, the rise in cereal prices is likely to result in a greater cereal acreage in wetter parts of the UK that are now under grass. In World War II such an expansion was achieved without benefit, generally, of meteorological predictions and many mistakes were made. We do not want to have to re-learn all this the hard way. Perhaps our agronomists, soil scientists, and meteorologists should get together now and predict the likely mean yield, given reasonable husbandry, for districts outside established cereal areas, and, more important, the number of years in ten when yields would probably fall below, say, 75 per cent of the mean.

On a world scale, future land use may present major problems as population grows. In general, not only the means, but also the reliability of yields of crops and of grazing livestock decrease as one moves towards warm-dry or cold-wet or cold-dry climate, away from the hydroneutral centre of the climatic spectrum of temperate farming systems (Duckham and Masefield, 1970, Table 3.1.1, Fig. 3.1.11 and p 480, 482), ie, as the unreliability of the climate increases and as mean effective transpiration falls. A hydroneutral area is one where the potential evapo-transpiration approximately equals precipitation in the thermal or the hydrologic growing season. Hydroneutral areas, eg, US Corn Belt, countries round the North Sea, parts of New Zealand, tend to have a greater range of enterprise choice, higher industrial inputs per unit of effective transpiration, and higher yields per unit area than areas on the warmer, drier or colder, wetter sides of hydroneutrality (Duckham and Masefield, 1970, p 459 ff and p 481).

Thus, the greater the expanded output required to meet population and other pressures or to release 'good'* farm land for, eg, urban development, then the greater is the need either (a) to increase output per unit area of good land, or (b) to use areas which have not only lower effective transpiration but unreliable climates. At present we tend to rely on the former, ie, increased yields. But for technical (eg, loss of soil structure) or social (eg, pollution of inland waters by fertiliser residues) reasons there may in future be limits to this process. We may have to reverse present trends and increase output from 'marginal' areas.

If we do have to make more use, in the temperate world, of unreliable climates then forecasts, especially of reliability (on the lines of Hogg's, 1967, frequency of irrigation-needs), will become increasingly important in marginal areas. This is firstly because there are, in such areas, usually fewer 'buffers' (silage stocks, grain stores, etc) and secondly because farmers tend to over-estimate the risks and to inject fewer inputs than meteorological

---

* Land which is fairly flat, has workable soil and a hydroneutral climate with an effective transpiration over 500 mm per year.

probabilities justify (Duckham, 1963, p 190; Duckham and Masefield, 1970, p 481). Anything that helps to give the farmers (and their sources of agricultural credit) justifiable confidence helps to raise biological, and usually economic efficiency. But in presenting their conclusions, meteorologists are usually wise enough to be cautious. Over-confidence (eg, in the feasibility of maize production in western Nebraska, USA) can lead to a generation of under-confidence.

In the tropics, especially in the vast seasonally dry areas, the situation is more difficult. Farming is less sophisticated; climates are unreliable and there are fewer buffers, such as roads to bring in grain stocks, if the rains or the crops fail. Peasants therefore 'play safe'. For instance, they may rely on mixed cropping (inter-cropping) even if this practice precludes simple technical advances. They farm primarily for self-sufficient survival, even if they include a few cash crops for profit or to buy domestic essentials.

The meteorologist and the hydrologist have therefore key roles to play in helping tropical farmers to break out of the 'survival only' mentality. This arises because the dates, durations, and amounts of 'the rains' in areas of short, wet seasons (Wrigley, 1969, p 68) are at present unpredictable. The kind of field husbandry experiments used in temperate countries to test the feasibility of particular crops are, in many tropical areas, even if repeated over many years and at many sites, liable to give misleading results. 'Obviously', state Webster and Wilson (1966, p 17), 'data on the expectation or probability of rainfall throughout the season provide a much simpler, less laborious and more accurate means of assessing optimum planting dates.' For example, the meteorological work of Manning (1956, 1958) and his associates at Namulonge in Uganda showed that the probability of enough early rain justified bringing forward the traditional planting date and thence the harvest of maize. This then left enough time to plant and grow a cash crop, viz, cotton in the 'second' rains, and harvest it in the dry season.

Unfortunately, however, the lack of adequate rainfall records at enough sites and of data on the water requirements of tropical crops (Webster and Wilson, 1966, p 17) at present limit the predictive value of this approach. Nevertheless, the rainfall probability map of East Africa (Glover et al, 1954), estimates of drought risk, and its farming and human nutritional consequences, such as those now in hand in Tanzania (Kates and Wisner, 1971),

D

and possibly the use of 'short cut' weather data collection as in Israel (Duckham and Masefield, 1970, p 306) offer considerable predictive scope in the tropics, but only if their practical application can subsequently be demonstrated to the full satisfaction of the local farmers. In the tropics, more perhaps than in temperate areas with 'good' climates, the meteorologist must be a key member of the food production research team.

Investment in 'Weather-proofing' and Environmental Improvement

Long-term investments of this type increase fixed annual costs. Their analysis is not easy, partly because it is not often possible to separate weather-proofing fixed costs from other fixed costs, such as the rental value of the land, and partly because weather-proofing investments often require, to make them economical, added consequential investment, eg, in more livestock or in increased variable costs such as fertilisers. The following model, therefore, is crude, over-simplified, and has obvious defects; thus it omits the consequential added variable costs and the potentially increased output and profits springing from 'weather-proofing' or 'climate-improving' fixed costs. It is limited to operating costs.

Let

$J =$ Annual fixed costs needed to complete a particular job (eg, the combine harvesting and drying of grain) or the operation of a farming system when working under ideal conditions with, in each case, no weather-induced losses of productive time (W_a) or of yield (W_b).

W_a and $W_b =$ Weather-induced losses, if no counteraction is taken, of (W_a) productive time in any given year; and of

(W_b) yield losses (eg, due to shedding of grain or wild life predation during harvest, or high death rates in newly born outdoor pigs) in any given year.

$I =$ Extra annual fixed costs attributable to investment, in excess of that required for J, to eliminate W_a and/or W_b in n years out of ten.

E = Extra variable costs (other than the consequential costs mentioned above) which *either* increase the capacity of $J+I$ *or* replace all or part of I (eg, using more oil fuel to dry grain more rapidly at higher temperatures, or using additives such as propionic acid to reduce mould growth in wet stored grain, or the heating of pig houses).

Then if n is 7, that is, if W_a and/or W_b are to be eliminated in seven years out of ten, the model is:

$$\text{Mean annual operational costs} = \frac{10(J+I+E)+3[(W_a+W_b)-(I+E)]}{10} \tag{1}$$

To apply, usefully, a model of this type in order to decide how much to invest in weather-proofing, we need the help of the meteorologist or at least we need access to his records for particular localities and for months or days or even parts of days. Without such aid it is difficult to predict the cost of, and the return on, such investment.

We now apply the above to the three main forms of relevant investment, ie, water control, field equipment and machinery, and farm buildings.

Water Control
As in the case of current land use, many flood-control, land-drainage, and irrigation schemes have been installed without the benefit of meteorology and hydrology. But today predictions of potential or actual crop losses (W_b) due to flood risks, poor drainage or irrigation-needs and to lost biological time (W_a) (eg, absence of double cropping),* are increasingly taken into account when predicting the needed capacity of such water control and storage schemes and the potential improvement in edible or saleable crop or livestock yields.

Although in markedly dry climates the farmer's use of irrigation water is often based more on custom or his water allocation than on actual water need at the time, in many countries, eg, in the UK, the capital invested, *I*, in

---

* For an analysis of the food losses between receipt of solar radiation and precipitation and final food consumption see Duckham on *Human Food Chains* in Dent and Anderson (1971).

private sprinkler schemes on individual farms is increasingly determined by the probability of the frequency of water deficits and the size of crop response to irrigation water.

But, as noted above, this raises the problem of added consequential costs. Thus, one can predict that the capital cost of a sprinkler installation will be, say, £200 per acre and that the consequential investments in working (variable) and fixed capital (eg, in more fertiliser use, in doubling cows per acre of grass, in more cow housing, in a larger milking parlour, etc) may come to more per acre than the reservoir or borehole, pumps, pipes, and sprinklers. This problem of consequential investment may be critical in developing countries where the farmer has very low cash or credit resources. Thus, when first using water for irrigation from a public scheme, he ought, for example, to buy fertiliser to exploit the rise in the effective transpiration of his land. It is strange that the US Presidents' Science Advisory Committee (PSAC, 1967), which included a world survey of irrigation-needs and their capital requirements in its report on the *World Food Problem*, paid little consideration to the consequential annual fixed and variable costs needed to exploit watershed, drainage, and irrigation schemes.

Field Equipment and Machinery
Recent work by Dalton (1969), Gemmill (1969, 1971), van Kampen (1969), and van Kampen in Dent and Anderson (1971), has outmoded earlier suggestions by Duckham (1963, p 343, 1966), who failed to include yield losses (W_b) in his model. Field experiments and extrapolations by simulation models have shown that losses of yield due to weather-induced untimeliness in sowing or planting dates or in harvesting can be severe. Thus yield losses of 250 kg per hectare per week have been reported when drilling of spring cereals was delayed after mid-March (Stansfield, J. M., in Taylor, 1970), Gemmill (1969) estimates that, on average, untimeliness, ie, yield losses (W_b) for a 'model' arable farm of 120 hectares on heavy clay in north-west Essex, England, would, over his analysis of ten years' weather, exceed annual machinery costs except in exceptionally good operating years like 1959. Second, his work suggests that as mean machinery costs ($J+I$) increase from £600 to £900 per year, so mean untimeliness costs would fall from £1650 to £1250 per year. This hardly suggests that added fixed costs, I, to reduce un-

timeliness losses (W_b) are, on average, justified. But farmers cannot live on averages. They need an income each year, or at least their employees, bank managers and creditors expect them to have one. So it is deviation from the mean that is critical. In a bad operating year such as 1963, weather-induced crop losses (W_b) on Gemmill's data, would, with annual machinery costs ($J+I$) at £600 per year, be nearly twice the crop losses (W_b) predicted if machinery costs were £900 per year. In such a year the extra £300 (900 − 600) fixed costs would have reduced crop losses (W_b) by more than £1200.

This emphasises the need for meteorological, and consequential biological probability data on the frequency of 'good' and 'bad' years in helping farmers to decide how much to invest in weather-proofing ($I+E$) to offset weather-induced crop losses (W_b) and weather-induced losses of machine time and of biological time, eg, possible catch crops (W_a). (The problem is rather more complex, however, than the above example implies. For a 'good' operating year may coincide with a drought year, like 1969, when yields as well as weather losses are low and higher market prices may not compensate for lower biological output.)

Nevertheless, there is much scope for meteorologists to help in predicting the probabilities of 'good' and 'bad' operating weather as an aid to wise 'weather-proofing' investment in expensive field machinery designed to curtail weather-induced losses (W_a and W_b).

Farm Buildings

Farm buildings account for about 60 per cent of the total of over £600 million of long-term capital invested in British agriculture over the last decade (Harvey, 1970). Whilst recent investment in farm buildings may be relatively less in western Europe and USA and much less in, eg, Australia and New Zealand, it is clear that the capital 'trapped' in farm buildings is very substantial in the temperate world. Farm buildings facilitate biological control (for example, segregation by age, sex), ease work organisation (as in milking procedures), provide organised storage, save pastures from 'poaching' by cattle hooves, reduce loss from human and other predators, provide comfortable work conditions for man, and, finally, protect livestock from inclement weather so as to reduce weather costs, ie, losses of yield due to adverse weather (W_b). Most of these functions in reality need only a roof and some wire-netting. It is the last (livestock protection against weather) that is, on most farms in cool and sometimes in seasonally hot climates (eg, Israel), expensive and probably less necessary than it is usually assumed to be. Thus at Sonning Farm, University of Reading, in 1966 the capital tied up in livestock protection was about £75 sterling per acre (Duckham, 1966).

A simplified energy model for livestock production (Duckham, 1966) is:

$$Y = f[C - (M_b + M_c + M_d)] \qquad (2)$$

where Y is yield, C is the metabolisable energy in the intake of the animal, M_b is the basic metabolism of the resting animal in thermoneutrality, M_c is the energy need attributable to non-thermoneutral conditions in the animal system, and M_d is the energy need attributable to the acts of grazing or seeking shelter or shade from bad weather conditions. Factors which reduce metabolisable intake, C, or increase weather-sensitive maintenance needs (M_c, M_d) reduce productive yield. By housing livestock properly we reduce the adverse effects of hot or very cold weather on consumption (C) and/or of cold-chill, wind-chill, and rain-chill on the energy cost of maintaining body temperature (M_c). But how frequently, how much, and in what seasons of the year are heat losses due to chill so raised, or appetite depressions due to excess heat so great, that they justify expensive housing?

Meteorologically the problem is roughly the same as with field machinery. Assume that the non-weather-sensitive fixed costs (eg, segregation of livestock by age, etc) are J and irreducible. Then, how much can the farmer usefully invest in 'weather-proofing' farm buildings (I) and on feeding or heating (E) to offset probable weather-induced losses of productive time (W_a) or, more important, loss of yield (W_b), ie, how much should he invest to reduce losses due to lower C and higher M_c in model (2)? The answer broadly is that we do not know. We may know something about (a) the insulating capacity of buildings and flooring materials, (b) the heat produced by animals of different species, sizes, and at different dietary levels, (c) ventilation needs of animals, (d) the probabilities of high or low temperatures, etc, and (e) micro-climates within buildings. But a synthesis of this complex into, for instance, a simulation model does not seem to be available. For the present, therefore, the role of the meteorologist in helping to predict what level of fixed (I) or variable (E) costs are justified by the weather probabilities seems to be limited.

But buildings are, as suggested above, at least in many parts of north-west Europe and North America, one of the major long-term investments trapped in agriculture and much, or perhaps most, of this investment is for protecting livestock against weather. There is, therefore, a strong case for a combined attack on this problem by farm buildings experts, animal physiologists, and meteorologists.

Conclusions

For the management of agriculture, whether at the farm, local-merchant, local-banker, or government levels, medium-term, seasonal, bio-meteorological forecasts can be helpful, whilst, for decisions on land-use and enterprise choice and fixed capital investment, probability data on a wide range of weather parameters are badly needed. Such probability data ought to enable reasonable forecasts to be made of the outcome of particular decisions (eg, to extend the farming area, to introduce new crops, to install irrigation, to house livestock, to reduce or to increase the safety margin in the combine-harvester capacity). This applies especially in the tropics where, however, short-cut meteorological devices may have to be devised to predict (a) the feasibility of particular crops or farming systems, (b) the need for irrigation or flood control, or (c) the risks of drought or famine.

In using weather probability data to assess the validity of climate-modifying or weather-proofing investment, the problem of the consequential fixed and variable costs that are needed to exploit the basic investment (eg, in irrigation works) must not be overlooked.

Finally, just as hospital medicine is increasingly team work, so is agricultural progress; the meteorologist must become, especially in the tropics, a full member of the food production team.

Acknowledgements

Thanks for information are due to L. P. Smith, Meteorological Office, and to E. H. Roberts, J. Pearce, and J. M. Stansfield, Department of Agriculture, University of Reading.

References

DALTON, G. E. (1969). *Methods for Improving Investment Decisions in Farming*. PhD Thesis, University of Reading.

DENT, J. B. and ANDERSON, J. R. [Eds] (1971). *Systems Analysis in Agricultural Management.* Wiley (Australasia), Sydney, New York, and London.

DUCKHAM, A. N. (1963). *Agricultural Synthesis: The Farming Year.* Chatto & Windus, London.

DUCKHAM, A. N. (1966). *The Role of Agricultural Meteorology in Capital Investment Decisions in Farming.* University of Reading, Department of Agriculture, Study No. 2.

DUCKHAM, A. N. and MASEFIELD, G. B. (1970). *Farming Systems of the World.* Chatto & Windus, London: Praeger, New York.

GEMMILL, G. T. (1969). *Approaches to the Problem of Machinery Selection.* MSc Thesis, University of Reading.

GEMMILL, G. T. (1971). *Personal communication.*

GLOVER, J., ROBINSON, H. C., and HENDERSON, J. P. (1954). Provisional Maps of Reliability of Annual Rainfall in East Africa. *Quart. J.Roy. Met. Soc.,* **80,** 602.

HARVEY, N. (1970). *A History of Farm Buildings in England and Wales.* David & Charles, Newton Abbot.

HOGG, W. H. (1967). *Atlas of Long-Term Irrigation Needs for England and Wales.* Ministry of Agriculture, Fisheries & Food, London.

KAMPEN, J. H. VAN (1969). *Optimising Harvesting Operations on a Large Scale Grain Farm.* PhD Thesis, University of Wageningen, Netherlands (quoted by Dalton and by Gemmill above).

KATES, R. W. and WISNER, B. (1971). *Personal communication.* Clark University, Mass., USA.

MANNING, H. L. (1956). The Statistical Assessment of Rainfall Probability and its Application to Uganda Agriculture. *Proc. Roy. Soc. B.* **144,** 460 (quoted by Webster and Wilson, qv).

MANNING, H. L. (1958). The Relationship between Soil Moisture and Yield Variance in Cotton. *Rep. Conf. Directors etc. of Overseas Dept.* Col. Office Misc. **531,** p 47. HMSO, London (quoted by Webster and Wilson, qv).

PSAC (1967). *The World Food Problem.* US Presidents' Science Advisory Committee, Washington DC (3 vols).

ROBERTS, E. H. (1970). *Personal communication.* Department of Agriculture, University of Reading.

SMITH, L. P. (1968). Forecasting Annual Milk Yields. *Agric. Met.* **5,** 209–14.

SMITH, L. P. (1970). *Personal communications.*

TAYLOR, J. A. [Ed] (1970). *Weather Economics.* Pergamon Press, Oxford and New York.

WEBSTER, C. C. and WILSON, P. N. (1966). *Agriculture in the Tropics.* Longmans, London.

WOOLHOUSE, H. W. [Ed] (1969). *Dormancy and Survival.* Symposia of the Society of Experimental Biology. Cambridge University Press, Cambridge.

WRIGLEY, G. (1969). *Tropical Agriculture: the Development of Production.* Faber, London.

CHAPTER SIX
W. H. HOGG

The Weather Forecasting Requirements of Specific Types of Agriculture and Horticulture

Introduction

In this chapter an attempt will be made to assess the extent to which weather forecasts are used by the farming community, how they are used, the ways in which they could be improved, and the demand for other types of forecast. The assessments are based on the answers to questionnaires distributed to selected farmers in Wales and the South West and West Midland Regions (in England) of the Agricultural Development and Advisory Service. The sample is by no means random, as the County Agricultural Advisers selected farmers who were likely to reply. Within each county the selection covered as wide a range of different enterprises as possible but the large-scale arable farming of eastern England is not represented, nor are the major horticultural areas of eastern and south-east England.

A copy of the questionnaire, somewhat modified to save space, is given as an appendix (p 82). As will be seen from the introduction, it is assumed that the forecasts are substantially accurate in order to concentrate on the question: 'What does a farmer need from a weather forecast?' Without this assumption there is little doubt that the primary requirement would have been greater reliability.

The response to the questionnaire was good. Some 60 per cent of the forms were returned. A few could not be used for various reasons, and the final numbers used in the analysis were as shown in Table 6.1.

Table 6.1

RESPONSE TO QUESTIONNAIRE

Region	Circulated	Usable replies
South West	85	51 (60%)
West Midland	76	45 (59%)
Wales	126	70 (56%)
Total	287	166 (58%)

Daily Forecasts: Present Use

The extent to which the existing 24–48 hour forecasts are used can be obtained from the answers to questions A18–A24 (p 83). These clearly show (A18) that a large majority of those questioned use forecasts in planning their work, often or regularly, with a marked emphasis on the period April to September (A19), see Table 6.2.

Table 6.2

PERCENTAGE OF FARMERS USING FORECASTS

Regularly (*R*), Often (*O*), Seldom (*S*), and Never (*N*) and percentage of entries for winter (*O–M*) and summer (*A–S*) half-years.

	R	O	S	N	O–M	A–S
South West	57	39	2	2	28	72
West Midlands	62	33	5	0	25	75
Wales	39	49	11	1	20	80
Total	50	42	7	1	24	76

The forecasts for the winter half-year are often needed for work which is not urgent (cultivations and muck spreading are good examples), but specialists such as vegetable growers in the early areas of the South West use forecasts in relation to harvest, and sheep farmers may be vitally concerned at lambing. Most of the interest in summer forecasts relates to harvests and conservation, including silage-making, but certain other operations such as spraying are also highly weather-dependent.

Radio and television forecasts are by far the most used by farmers (with little difference between them) and these are followed by forecasts obtained by telephoning a Meteorological Office (A20). Little use is made of forecasts in the press or recorded forecasts and an analysis of preferred times showed a marked wish for forecasts around breakfast-time, lunch-time, early evening, and late evening (A21) (Table 6.3).

Table 6.3

FORECASTS USED AND PREFERRED TIMES

	TV	R	N	R(T)	F(T)		B	L	E_1	E_2	X
			Types						*Times*		
					per cent						
South West	32	43	4	6	15		36	17	13	20	14
West Midlands	36	41	1	3	19		37	19	13	15	16
Wales	47	33	7	1	12		30	22	10	23	15
Total	39	39	4	3	15		34	19	12	20	15

Note: The headings for the types of forecast are abbreviations of the items in question A20. The headings for the times are: B, Breakfast, 0630–0825; L, Lunch, 1230–1325; E_1, Early evening, 1730–1825; E_2, Late evening, 2030–2225; X, all other times.

The percentages are based on the number of preferences expressed and not on the number of usable returns.

Exactly 50 per cent of the Welsh farmers stated that they telephone a Forecast Office, compared with 84 per cent for the West Midlands and 76 per cent for the South West. This regional difference is no doubt due to the greater distances and costs in Wales, with perhaps a feeling that a forecast made at a distance will not be sufficiently accurate for decisions on a particular farm

(A22). These forecasts are used mainly during May–September (80–90 per cent, according to region). They are used for a variety of farming activities—particularly for the spraying, conservation, and the harvesting of a wide range of crops, eg, strawberry picking, hop picking, and herbage seed harvest. In winter, the harvest of such crops as leeks and Brussels sprouts can be affected by weather. Lambing was mentioned as another reason for using forecasts in winter (A23). When using general forecasts, well over 75 per cent of farmers attempt to modify them on their own local knowledge and experience (A24). Farmers are about equally divided on the question of including references to farm operations in weather forecasts (A25)—a large majority in Wales favours this but a large majority in the West Midlands is against.

In questions A1 to A13, farmers were asked to state whether individual elements were important to them. The percentages of affirmative answers are given in the first four lines of Table 6.4, based on the numbers of usable replies (see Table 6.1). The reasons for which the information was likely to be used are given in the bulk of the same Table 6.4, each mention of the particular farm operation being given a score of 1. Information on sunshine and rainfall amount is wanted by the largest number of farmers, mainly in relation to silage-making, hay-making, and the cereal harvest. Rainfall information is also useful in relation to a number of other operations, particularly cultivation and ploughing, drilling, sowing and planting, fertiliser application, spraying, and stock management. Frost figures high, for obvious reasons, followed by general windiness, a major factor in spraying, hay-making, and the cereal harvest. The importance of snowfall is emphasised by stockmen and shepherds and, less frequently, by dairy farmers who must get their milk to collecting points. Only four elements were considered unimportant by the majority of farmers: night temperatures and frozen ground (perhaps because they are largely satisfied by the information on frost), hailstorms, apparently the least important, and gales as opposed to general windiness, though these showed up as important in relation to any building construction and maintenance.

Questions were asked on combinations of elements, and wind frequently appears here. The two most important combinations are wind and rain, and wind and sunshine (A14, A15). Wind and rain affects the management of stock and a number of operations such as spraying, silage and hay-making,

Table 6.4

IMPORTANCE OF ELEMENTS (PER CENT) AND THE ASSOCIATED FARM
OPERATIONS (FREQUENCY)

	A1	A2	A3	A4	A5	A6	A7	A8	A9	A10	A11	A12	A13
Regions							*per cent*						
SW	65	59	47	71	45	84	92	69	73	24	76	49	73
WM	62	64	47	82	36	82	80	56	53	18	69	31	69
Wales	61	44	37	67	31	83	76	46	70	14	63	57	46
All	63	54	43	72	37	83	82	55	66	18	69	48	60
Operations							*frequency*						
CP	4			9	22	2	12	6	2		2		2
D	16	6	3	12	5	4	23	11	2	1	10	2	1
FA	19	3	1	3	5	4	10	5	1		12	4	4
S	21	14	5	3		5	15	9		1	63	11	14
I		1				1	7	1			1		
Sil	7	8				29	26	11			9	2	8
HM	28	31	3			98	54	26		1	31	6	47
CH	21	15	1			62	39	25		3	23	13	34
RH	5	2		7		2	5	1			1		1
HSH	3	2				1				1	2	1	3
VH	2	1	2	5	1	2	2	1	3		1		1
FrH	3	2				1	2					5	1
FlH		1	1	1	1	1	2	1	1		1		1
CD		2	3										20
CS	3		9	16									6
SM	27	22	21	19	5	2	20	14	63	3	2	10	4
H	2	2	4	3							1		2
FP				38	1								
FC				16									
GM	3	2	4	3		3		2	1	2		1	1
BCM				3								18	
T				3	20		3		28				
DW					1		2	1					
HC					1					1	1	1	
GPR									1	1	1	2	
HW	1	1	1	1			1	1		1	1	1	
M	1								1				

Key: CP Cultivation and ploughing: D Drilling, sowing, and planting: FA Fertiliser application: S Spraying: I Irrigation: Sil Silage-making: HM Hay-making: CH Cereal harvest: RH Root harvest: HSH Herbage seed harvest: VH Vegetable harvest: FrH Fruit harvest: FlH Flower harvest: CD Crop drying: CS Crop storage: SM Stock management: H Housing of animals: FP Frost protection (pipes): FC Frost protection (crops): GM Glasshouse management: BCM Building construction and maintenance: T Transport: DW Drainage work: HC Hedge cutting: GPR Glasshouse protection and repairs: HW Hop harvest and maintenance: M Marketing.

and combining; sunshine combined with wind (perhaps air movement would be a better term here) favours hay-making, corn drying, and spraying, though calm weather is best for the actual spraying with some breeze later to dry the applied spray. Snow drifting produces many complications for the hill farmers concerned largely with livestock, and they are alert to forecasts of wind and snow. The only treble combination mentioned is temperature, sunshine and humidity, in relation to hay-making. For many of the operations mentioned above the farmers are primarily interested in the 'drying power' of the air. Only 25 per cent of the farmers thought other elements were important. The two most frequently mentioned were soil temperature and visibility (A16, A17). The mention of visibility was a little surprising, though clearly the movement of stock could be affected, and also ploughing and drilling.

Daily Forecasts: Satisfaction and Shortcomings

Questions A26–A33 deal broadly with the satisfaction with the present forecasts, again excluding the factor of accuracy. The answers for these are summarised in Table 6.5, with further notes below.

Table 6.5

PERCENTAGE OF AFFIRMATIVE ANSWERS TO
SELECTED QUESTIONS

	A26	A27	A28	A29	A31
South West	92	43	57	80	88
West Midlands	98	40	47	69	87
Wales	93	50	59	76	80
Total	94	45	55	75	84

There is a great demand for information on the probability of rainfall (A26). Opinion is about equally divided on the detail provided on TV (A27) and radio (A28) broadcasts, although a considerable majority of farmers think there are enough TV (A29) and radio (A31) broadcasts. For TV, the most favoured times for additional forecasts were: 0630–0825, 1230–1325, 1730–

1825, and 2030–2225; see also Table 6.3. For many farmers the lunch-time forecast is rather too late except on Saturdays when it is brought forward, and is in fact too early for some farmers. The request for additional forecasts at 1730–1825 is surprising in view of the present timing.

Daily Forecasts: General Comments
In question A33 farmers were asked for additional comments which would help to explain what they think forecasts should contain. It is not possible to summarise these comments on a numerical basis, and the following notes indicate the points which were most frequently mentioned.

1   The need for more local forecasts. The difficulty of providing these was understood by most farmers, some of whom suggested that they should be broadcast by local radio stations. Perhaps the most ambitious suggestion was for hourly forecasts.

2   The importance of Atlantic charts on TV was stressed and some farmers asked that information on the movement of systems should be given as fully as possible, clearly so that they could make some estimates of their own.

3   The importance of the individual weather elements has been dealt with earlier, but many farmers emphasised rainfall. If this could be more accurately forecast (timing, duration, and intensity) a large range of farming operations could be more efficiently planned and undertaken. With hay-making and harvesting in mind, a number of farmers asked for more information on drying weather.

4   It is clearly important for us to accept that many farm operations are spread over several days and that the outlook should possibly cater for this, even though vague at the end. Certainly some obviously successful farmers are prepared to accept this as a basis for general planning, subject to a check with the nearest forecast office in relation to individual decisions.

5   Some points were raised which are not related to the technical content of a forecast but to difficulties in following the presentation. These included the lack of understanding of the meaning of weather symbols on TV charts and uncertainty as to location.

Weekly Forecasts

The only guidance at present given on a weekly basis is contained in the BBC Farming Programme on Sundays. The questionnaire did not specifically deal with this, but some farmers commented on it. Over 90 per cent of farmers thought that weekly forecasts would help them (B1) and the elements on which information is most required (B2) are, in descending order, rainfall, sunshine, temperature, humidity, snow, wind, and hail. The farming operations affected are similar to those mentioned earlier in relation to short-term forecasts: the importance of rainfall throughout the year was emphasised in many replies.

Two main points emerged from the answers to question B3 which asked for further remarks on the possible uses of weekly forecasts in farming. The first of these was the acceptance of their value for planning. The second was the interest in the weather prospects for farmers and growers which is included in the Sunday Farming Programme on BBC television. A few of the replies expressed some doubt as to their value, but far more commented favourably on them, often suggesting that they should be revised at least once during the week.

Monthly Forecasts

The monthly weather prospects which are now issued twice monthly are seen or heard by 73 per cent of farmers (C3) but only 31 per cent make any use of them in their farming (C4). Not many more, 36 per cent, thought that they would be helped by the inclusion of other weather elements (C1). Apart from temperature and rainfall, already included in the monthly weather prospects, there is substantial demand for sunshine and humidity (C2). Farmers who use the present information cite ten different farm operations in which it is applied (C5). Of the 52 farmers using it, 13 mention hay-making, 7 drilling, sowing, and planting, 6 cereal harvesting, 5 each silage-making and stock management. Others mentioned less frequently are cultivation and ploughing, fertiliser application, spraying, root harvesting, and transport.

If the monthly weather prospects were more accurate, 81 per cent consider that they would be an aid in farming (C6). Although the term precision was not used in the question, there may have been some confusion between precision and accuracy. Fifteen operations are thought likely to benefit from 'improved' monthly forecasts and of these the five most frequently mentioned are: hay-making, cereal harvesting, stock management, drilling, planting and sowing, and silage-making (C7).

Question C8 asked for ways in which accurate but general forecasts could help the farmer and, as with the weekly forecasts, the importance of planning was emphasised. A few farmers indicated that shorter-term forecasts up to about a week were more important to them. The importance of monthly forecasts in relation to grassland management is evident. In autumn a more open season will allow the stock to be kept out longer; in spring information on rainfall could lead to the saving of two weeks' winter feed by the application of fertiliser and controlled grazing.

The arable farmer is probably most interested in the timing of cultivation and drilling. He wants to know when he can safely get on to the land, and the generalised monthly forecast should be sufficiently accurate for this—it should enable him to make his own estimates of the soil moisture status and soil temperature a few weeks ahead.

Forecasts needed for Specific Types of Farming
To complete this summary of the survey, a few of the 166 usable replies have been selected and are described in some detail to indicate the forecast requirements for different types of farming. These cannot cover the whole range of agriculture and horticulture but are suggestive of the general needs in the western districts of England and in Wales.

Livestock Farm
This is a farm of some 200 hectares in Cardiganshire where fat lamb production and store cattle production are the main enterprises. The farmer uses radio forecasts in planning his work; he finds there are enough of these but would like more local forecasts—he does not live near a Forecast Office. The main uses of daily weather forecasts to this farmer are given in the table on p 78.

This farmer would like weekly forecasts (he has no television set and does not see the Sunday Farming Programme) and thinks he would use them for most of his farm operations, eg, for cultivations and harvesting in the spring and summer and for stock management in winter. The present monthly

Weather element	Period	Farm operation
General temperature	Jan–mid Mar	Supplementary feeding of sheep, especially in relation to snow cover.
	Late Feb–Mar	Fertiliser application must be timed to anticipate soil temperature for grass growth.
Day temperature	June–Sept	Hay and corn harvest (drying) (sunshine and humidity also relevant here).
Night temperature	Nov–Dec	Cattle housing.
Frost		Freezing of water supplies to stock.
Frozen ground	Feb–Mar	Ploughing and cultivation.
Rainfall		Amount affects stock housing. Intensity affects shearing.
Snowfall		Disruption of communications and transport affects shepherding and the outdoor feeding of stock.
Gales		General damage, including buildings.
Fog		Stock management in open areas.
Rain, Cold weather	Winter	Stock housing.
Wind, Snow	Winter	Blizzards–stock management.

weather prospects are of more interest than value to him but if they were more reliable he believes he could use them for timing fertiliser applications, and for rationing and housing his stock.

Milk Production
This Somerset farm covers about sixty hectares and the main enterprises are the production of milk and the rearing of dairy herd replacements. The farmer regularly uses radio forecasts, but never telephones a Forecast Office— the area in which he farms is not well served in this respect. He thinks there are sufficient TV and radio forecasts but asks for more local details on the radio.

He would welcome weekly forecasts especially during the silage and hay-making season. This also applies to accurate monthly forecasts, which he would also use for more economical use of fertiliser for early grazing.

This farmer is interested in the daily forecasts for the following reasons:

Weather element	Period	Farm operation
General temperature	Mar–Oct	Control of grazing, particularly when cattle are put out in spring.
Night temperature	Feb–May	Rearing of young calves, control of ventilation.
Frost	Nov–Mar	Anti-freeze precautions in relation to machine milking and drinking troughs.
Frozen ground	Dec–Feb	Manure spreading—provides opportunity to reduce accumulation during housing period.
Sunshine	May–Aug	Hay and silage-making, rapid wilting and drying.
Rainfall amount	May–Aug	Silage and hay-making, control of cutting policy.
	Mar–Oct	Grazing management, minimising poaching especially on newly sown leys.
	Mar–Sept	Fertiliser application.
Rainfall intensity	Mar–Oct	Flooding of Somerset Levels.
Snowfall	Mar–Apr	Decisions on cattle housing.
Humidity	May–Aug	Silage and hay-making.
Wind Rain Temperature	Mar–Apr	Grazing. If severe weather predicted, management modified to offer protection.

Arable farms

Replies were received from some largely arable farms in the east of the area under discussion. In order to cover a fairly wide range of crops the uses to which forecasts are put have been combined for six farms in Gloucestershire, Wiltshire, and Worcestershire. These are each some 240 to 360 hectares, and the crops include cereals, grass seed, beet seed, peas, beans, and potatoes. The main uses are as follows:

Weather element	Period	Farm operation
General temperature, day and night	July–Sept	Harvest, herbage seed and cereals.
	Aug–Jan	Harvest, potatoes and beet.
	Mar–May	Fertiliser application.
	April–May	Drilling and spraying.
Frost	Winter	Potato lifting. Cultivations. Potato storage (clamp protection).
Frozen ground	Winter	Ploughing and cultivation.
Sunshine	Mar–Nov	Bed preparation. Harvesting.
Rainfall		All land work.
	June–Aug	Cereal and grass seed harvest.
Snowfall		All land work. Access.
General windiness		Drying at harvest (with sun).
	Mar–Nov	Broadcasting, spraying, harvesting (also, gales).
Humidity	July–Sept	Harvest, cereals and herbage seed.
	Autumn	Grain drying in store. Potato blight spraying.
Snow and wind		Access.
Sun and wind	June–Sept	Drying standing corn. Harvesting in general.

All of these farmers think that weekly forecasts would help them, mainly in relation to fertiliser application, drilling, spraying, and harvesting. The same general views were expressed in relation to monthly forecasts.

Horticulture

Horticultural enterprises may be very specialised. One grower may concentrate on tomatoes and chrysanthemums under glass, another on top fruit, and a third on salad vegetables; their interest in forecasts will therefore vary considerably. In the area investigated, horticulture is not widespread and no attempt has been made to cover the whole field. As an example, details are given from a grower in Worcestershire with some thirty hectares of apples, pears, plums, and gooseberries and some additional land for vegetable production.

This grower is no doubt able to read more into forecasts than most people as he had some meteorological training. He regularly uses forecasts, both TV and radio, and telephones the local Forecast Office at any time of the year—in spring and summer when spraying and drilling, and in autumn and winter when frost may interfere with the harvesting of sprouts and leeks. He is satisfied that the present TV and radio forecasts are as frequent and as detailed as can be expected and that more local forecasts can be provided only by local radio stations.

Weather element	Period	Farm operation
Day temperature	June–July	Spraying top fruit. Knowledge of high temperatures important.
Frost	Dec–Mar	Vegetable harvest. Helps to determine time of starting work next day.
Frozen ground	Winter	Draining water from vegetable-washing machines and supply pipes.
Rainfall	Mar–Aug	Irrigation. Spraying. Cultivation.
Wind	Mar–Aug	Spraying.

He is very doubtful whether weekly or monthly forecasts would help him. This is interesting and against the general opinion expressed in the replies. Perhaps he has less need to get on to his fruit land early in the year, finds drilling and cultivation unnecessary, and can use chemical means of weed control.

Acknowledgements

This paper is published by permission of the Director-General, Meteorological Office.

Thanks are due to the County Agricultural Advisers in Wales and the South West and West Midland Regions of the Agricultural Development and Advisory Service; to the farmers who completed the questionnaires; and to my colleagues J. W. Davies and A. J. Pearson for assistance in sorting the information provided by the farmers.

Appendix: Questionnaire to Selected Farmers and Growers

Introduction

Many farmers use weather forecasts as an aid to management especially at certain times of the year, for example at hay-making or harvesting. The first requirement of a forecast is that it shall be substantially correct. Another important requirement is that the farmer is given the information which is most likely to be useful to him. In this survey we are assuming that the forecasts are accurate, in order to concentrate on the question: 'What does a farmer need from a weather forecast?'. Nobody is better able to answer this question than the farmer himself and this is why we are seeking your co-operation.

The questionnaire is arranged in groups according to the type of forecast. In each of these groups there are a number of questions to be briefly answered. We fully realise that it may not be easy to express some aspects of your requirements by answering these questions and we would welcome additional remarks.

GROUP A   24–48 hr FORECASTS   Questions A1–A33

The normal forecasts now issued cover a period up to 24 hours with an outlook for another 24 hours, sometimes longer. As we wish to find out what you want from forecasts, your entries will deal with the ways you use forecasts and the ways in which you would like to use them if they were more detailed or contained other information.

WEATHER ELEMENTS   Questions A1–A13

The list in column 1 contains the elements most likely to be mentioned in a forecast. For each one of these please decide whether it is important for you to know about it in advance. If so, please tick in column 2 (headed *Yes*): if not, please tick in column 3 (headed *No*).

If the answer is *Yes* please continue along the line, stating the farming operation the weather could affect and the months during which it is important. Any weather element may affect more than one farming operation and there may be several entries in columns 4 and 5 for one element.

1	2	3	4	5	6
	Important		Farming operation	Months	Remarks
	Yes	*No*			
	(tick in col 2				
	or col 3)				

WEATHER ELEMENT
A 1 General temperature
A 2 Day temperature
A 3 Night temperature
A 4 Frost
A 5 Frozen ground
A 6 Sunshine
A 7 Rainfall amount
A 8 Rainfall intensity
A 9 Snowfall
A 10 Hailstorms
A 11 General windiness
A 12 Gales
A 13 Humidity of air

COMBINATIONS OF ELEMENTS   Questions A14–A15
A 14 Are any combinations of elements (for example, rain and wind) important?
      Yes/No
A 15 If Yes, please list them with details: Combined elements; Farming operation;
      Months; Remarks

OTHER ELEMENTS   Questions A16–A17
A 16 Do you think any other elements not already mentioned are important?
      Yes/No
A 17 If Yes, please list them with details: Other elements; Farming operation;
      Months; Remarks

GENERAL   Questions A18–A33
A 18 How much do you use forecasts in planning your work? Please tick one of
      these: Regularly; Often; Seldom; Never
A 19 If regularly or often, please state months when they are most used and for
      what purposes:
A 20 If you use forecasts at all, which do you use most? Please tick one of these:
      TV forecasts; Radio forecasts; Newspaper forecasts; Recorded forecasts
      (telephone); Forecast Office (telephone).
A 21 At what times of day would you most like forecasts to be broadcast? Please
      give the time (or times) to the nearest hour, eg, 7 am, 1 pm, 7 pm
A 22 Do you ever telephone a Forecast Office?   Yes/No
A 23 If Yes, please state when (months) and why (farming operation):
A 24 Do you make any attempt to modify the general forecast by your local
      knowledge and experience?   Yes/No

A 25 Do you think that forecasts should include any references to the farming operations which may be affected by the weather which is forecast?   Yes/No

A 26 Would you like to have some indication of the probability of rain?   Yes (probabilities useful)/No
Example: There is a 60 per cent chance of rain today and 20 per cent chance tomorrow.

A 27 Do you consider that the present TV forecasts are sufficiently detailed? Yes/No

A 28 Do you consider that the present radio forecasts are sufficiently detailed? Yes/No

A 29 Do you think there are enough TV forecasts?   Yes/No

A 30 If not, at what hours would you like additional forecasts?   Times:

A 31 Do you think there are enough radio forecasts?   Yes/No

A 32 If not, at what hours would you like additional forecasts?   Times:

A 33 Additional comments: please add here any additional comments which will help to explain what you think forecasts should contain.

GROUP B   WEEKLY FORECASTS   Questions B1–B3

B 1 If weekly forecasts were available do you think they would help in your farming?   Yes/No

B 2 If Yes, which elements of the following would be most important, and how and when?   Temperature; Sunshine; Rainfall; Snow; Hail; Humidity of air.

B 3 Additional comments: please add here any further remarks you would care to make on the possible uses of weekly forecasts in farming.

GROUP C   MONTHLY FORECASTS   Questions C1–C8

Monthly weather prospects are issued at the beginning and middle of each month. These give in general terms the type of weather expected over about the following four weeks. They indicate whether the weather is expected to be warmer or cooler than average, and wetter or drier. The monthly weather prospects are published in the press and issued on TV and radio.

C 1 If monthly weather prospects included other elements, do you think they would help you in your farming?   Yes/No

C 2 If Yes, which elements of the following would be most important, and how and when?   Temperature; Sunshine; Rainfall; Snow; Hail; Humidity of air.

C 3 Do you regularly see or hear the monthly weather prospects?   Yes/No

C 4 Do you make any use of the monthly weather prospects in farming?   Yes/No

C 5 If Yes, please state how (farming operation) and when (month):

C 6 If the monthly weather prospects become more accurate, do you consider they may be an aid in farming?   Yes/No

C 7 If Yes, please state ways in which the prospects may help: Farming Operation/
       Months/Remarks
C 8 Additional comments: please add here any additional comments on the ways
       in which accurate (but general) monthly forecasts could help the farmer.

FINAL QUESTION
If it seems that a personal visit would further the object of this survey, would you
welcome this or not? Yes/No

The Value and Relevance of Weather Study and Weather Forecasting in the Profitable Production of Early Potatoes

The potato plant was first introduced into Europe about 1570. It is a native of the high, sunny, and relatively dry plateaux of the South American Andes, and is now widely cultivated as one of the staple foods of Europe and North America. Its successful cultivation is due, in large part, to the development of new varieties suitable to the British climate. Even so, on account of its origin and nature, it is still very sensitive to the rigours of this new environment.

Although a continuous supply of moisture is essential to the potato plant, heavy rainfall tends to cement the soil, thereby restricting aeration and tuber development. It is well suited to the dry, sunny areas of eastern England, particularly on deep, well drained land, rich in decaying vegetable matter. When grown in areas of high rainfall and humidity, it is very susceptible to attacks by various fungi, of which the most prevalent and destructive is potato blight, which caused the great Irish famine of the mid-1840s. It is particularly vulnerable to strong winds, frost, and very high temperatures. For the best results, it should be planted on deep, sandy loams. Conditions which seriously affect the keeping quality of the tubers are:

1 badly drained, heavy soils, liable to become waterlogged.
2 lifting and storing in wet conditions.
3 lifting for winter storage too early in the autumn.
4 too high a temperature in the clamps during storage.
5 inadequate protection against frost and dampness during storage.

It is no wonder, therefore, that the profitable production of early potatoes is, to such a large extent, dependent on the weather 'consciousness' of the grower, and that he should leave no stone unturned in his efforts to obtain and study all possible information concerning it. If he is fortunate enough to have received tuition in meteorology at school, and to have followed this up by a continued study of this fascinating subject, he should be well poised to absorb all the new technical data which meteorologists are now providing. At the beginning of this century, foreknowledge of the weather was largely based on observations, signs, old sayings, and barometers. Thus the forecasts, even in the short-term, were vague and unreliable. In recent years, however, great strides have been made in the preparation and presentation of forecasts. A mass of information, covering a very large area, has become available through the use of satellites, computers, wireless and television, etc. This will eventually diminish the role of the experienced meteorologist in preparing forecasts. It is to be hoped, however, that in the interest of continued progress a sensible balance will always be maintained between man and machine.

Reliability in weather forecasting is essential. Its successful application in the economic and commercial sphere, and the benefits derived therefrom, is the criterion by which its usefulness will be judged and its existence justified. Though short-term forecasts are now generally recognised as dependable and easily understood, making vital decisions on the strength of the accuracy of mid- and long-term forecasts is a much more complicated matter. This is where a knowledge of meteorology and a personal study of local weather conditions over a long period of years can be of great assistance. It is in this fusion of personal knowledge with all the available meteorological information, and the effective application of the resulting decisions, that the secret of success of the early potato grower lies.

Climate
Influenced as it is by the tropical zone of perpetual high pressure to the south, the dry, cold polar regions to the north, the warm currents of the Atlantic Ocean to the west, and the great land mass of Eurasia to the east, it is no wonder that the climate of the British Isles is variable, to say the least, and the weather very unpredictable. The sea, which is in constant movement, absorbs the sun's heat more slowly and to a greater depth than land areas, and retains

it for a longer period. Sea areas are consequently warmer in winter and cooler in summer than land areas. The prevailing south-westerly winds, relatively warm and laden with moisture, meet the cold north-easterlies from the polar regions, giving rise to low pressure areas on the Atlantic. These depressions are fairly consistent in their eastward movement, producing the mild wet British climate. Occasionally a ridge of high pressure develops between these depressions, providing a temporary fine spell, but it is only when pressure is high over the British Isles and Western Europe that long, dry periods of settled weather can be expected. It then becomes very cold in winter and warm and dry in summer. From the viewpoint of an observer in Britain, the sun is high and strong in late June and low and weak in December, thus forming the seasons. All this gives rise to an annual general pattern of anticipated weather which could be described as 'standard', with cold winters and warm summers. From time to time, however, for unknown reasons, the standard weather breaks down and abnormalities occur, often almost reversing the seasons, and giving winters with no frosts and summers without heat.

Areas Best Suited Climatically for the Production of Early Potatoes
The maps of the British Isles showing mean monthly isothermal lines indicate that south-western areas have the warmest winter and spring. This is conducive to earlier spring growth. The risk of a late spring frost is also much reduced. If the area is a peninsula, like Pembrokeshire or Cornwall, so much the better. The selection of a venue within such areas is also important if the best results are to be obtained. It is natural that the coastal fringes have milder winters and cooler summers than inland areas. The spring-frost risk is also less. Onshore breezes during hot weather in early June keep the temperature down, which assists earlier tuber formation. Complete freedom from spring frosts is more likely near a coast where a north wind comes straight off the sea. The earliest potatoes are produced on steep, sheltered, south-facing slopes, which are usually frost-free.

Standard Weather Pattern Expected Annually in the British Isles.
After a comparatively dry, anticyclonic period during September, winter rains usually commence in early October, these becoming heavier and more

prolonged during November, with stormy periods. Early in December a change to anticyclonic conditions can be expected with calm, cloudy weather or else cold fogs. From Christmas to the early part of February, it is likely to be cold and unsettled, with, occasionally, periods of frost and snow. It is not surprising, therefore, that after a long, cold, wet period, with hardly any evaporation, the land is usually too wet for cultivation during February. March heralds cold, drying winds and anticyclonic conditions once more, to be followed by cool, showery weather during April. Good growing weather, with wet periods intermixed with sunny spells during May, give way to a mainly dry, warm June. The expected weather for July and August can best be described as warm, close, and variable, with always a risk of thundery conditions developing.

The Ideal Type of Season for Early Potato Production
Standard weather, with a cold, dry March, wet May, and hot June, is not considered ideal for early potato production, because this leads to delayed lifting dates and heavy June–July crops. Plentiful supplies coming on the market from all early areas simultaneously cause disorganisation and uneconomic prices. This was the general trend during the 1960s. The early grower prefers a mild, unsettled winter to encourage steady growth during the chitting stage and to keep soil temperatures high. From the end of January until early March, it should be dry, in order to plant the seed in good soil conditions, the temperature being sufficiently high to promote root growth in early February. A moist, warm March and April induces early haulm growth, and a relatively dry May, with plenty of sunshine, improves the quality and accelerates the maturity of the new potatoes. June should be cold and showery with winds from a northerly direction, caused by an anticyclone over Greenland and low pressure over Western Europe. This should also check the spread of any outbreaks of potato blight which could be expected in such a season. On the other hand, the standard type of weather should be ideal for the production of seed potatoes. These are not planted until early April. A cold winter should prevent too much growth before they are planted, and a wet May should ensure sufficient haulm growth and heavy crops of seed-size tubers. In such a season it is usually possible to destroy the haulms of seed crops in early July, before they are attacked by blight.

General Survey of Weather Conditions in Pembrokeshire (1917–70)
(see Fig. 7.1)

The first half of this period was comparatively warm, with higher than average rainfall, the predominant south-westerlies producing the long and exceptionally moist growing seasons of the mid-1920s and 1930s. This was more or less repeated during the mid-1940s, which were ideal for early crops until 1949 (1947 being the exception). After 1949 conditions have been cooler generally, particularly during the 1960s.

If the period is examined in greater detail, it will be found that, *broadly speaking*, five short periods seem to conform with a standard weather pattern of cold winters and warm summers. They also conform with an *eleven-year cycle*. They are: (1917–21), (1928–32), (1939–43), (1950–54), (1961–5). The following cold winters occurred within these periods: 1917, 1929, 1940, 1941, 1942, 1963. The intervening years appear to have contained the greatest number of instances of unseasonable conditions and abnormalities. It is possible that this is due to the effect of increased sunspot activity at the time affecting our weather in a manner not yet understood. During these intervening years spring was sometimes very early, as in 1923, 1945, 1948, 1957, and 1960, or occasionally very late, as in 1947, 1955, and 1970. Very mild winters and cold, wet summers often occurred, as well as prolonged droughts in 1949, 1955, and 1959.

To note only a few freak occurrences within the intervening years, we may mention the 35·6 cm of snow falling overnight in Pembrokeshire, 1 April 1922; the severe frost and snow on 16 and 17 May 1935; the wet summers of 1935–6–7; the intense depression south of Iceland on 15 December 1945, when probably the lowest pressures ever recorded in these latitudes were registered by ss *Chinese Prince*, 928 mbs; the very dry November, 1945; the very cold, wet summer of 1946, rivalled only by 1903 and 1879; the amazing freak year of the century, 1947 when, in mid-January it was so mild that single sprouts on Arran Pilots grew an inch a day for six consecutive days; this was followed by two months of continuous hard frost and snow, and then widespread floods; then after a very unsettled midsummer, six consecutive weeks of cloudless skies; and a particularly warm spell in October–November.

Records on our farm in south Pembrokeshire over the last fifteen years

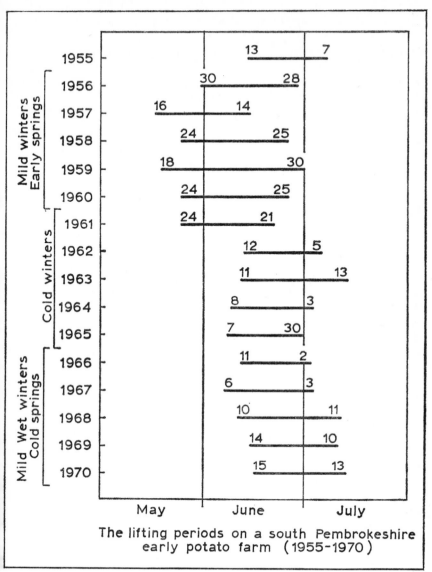

Fig. 7.1 Variations in the lifting periods on a south Pembrokeshire early potato farm (1955–70)

show that the intervening years (1955–61) were early lifting years for potatoes, but from 1961, cold winters and late springs delayed lifting by about three weeks, with consequent marketing problems. The expected milder winters returned between 1966 and 1970, but this time they were each followed by late springs and phenomenal mid-summer growth, this being prolonged into September and October. Assuming that the above trends will be repeated during the next two decades, we should expect a change before long to harder, drier winters and warmer summers; then more unreliable and abnormal conditions (1977–82), and much colder winters (1983–7).

Advantages Gained from the Study of Long-term Weather Prospects
Though the possibility that sunspots interfere with the weather is admitted by some meteorologists, there does not appear to be much enthusiasm for this theory, research having apparently proved so confusing and controversial that no use can be made of it in weather forecasting. It is an interesting study, however, to those who are endeavouring to explore new avenues. Decisions made on the results of such research should be considered purely speculative, but if the advantages to be gained far outweigh any disadvantages should things go wrong, then to an early potato grower in particular, the risk is worth taking.

The present series of four consecutive mild winters, with more than average rainfall, corresponding with the intervening years (1966–71) were anticipated on our farm and some decisions were made beforehand to try and reap any possible advantages. For instance, our seed potatoes, grown on our north Pembrokeshire farm (Farm Two in Fig. 7.2), were not lifted as early as in previous years, to prevent them getting too advanced in the chitting trays in a mild winter. As it happened, the late springs of those years also justified this policy and record crops were obtained on each occasion. Early-lifted home-grown seed, when planted in prolonged cold conditions following a mild winter, have a poor potential. They may even develop the condition known as 'little potato', ie, small new potatoes growing direct from the parent tuber, without haulm growth, but with ruinous results. Autumn ploughing and cultivations were also completed earlier, while the ground was dry. It was also decided to cut down on varieties which are known to be difficult to control in warm chitting house conditions.

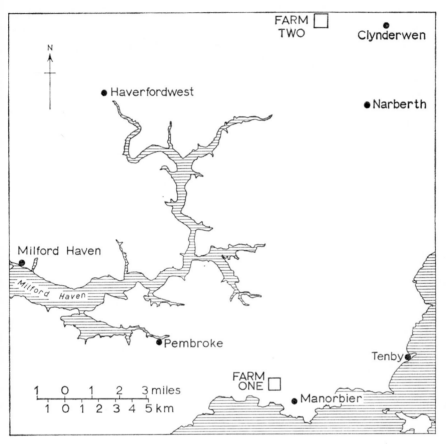

Fig. 7.2   Location of two Pembrokeshire farms

Advantages to be Gained from the Study of Short- and Mid-term Weather Prospects

Our seed potatoes are lifted during September and placed straight away into chitting trays, thereby avoiding the risk of overheating in clamps or heaps. A maximum temperature of 15·6°C is ample to heal any damage suffered during

E

lifting. Higher temperatures will cause premature germination and a temperature of 37·8°C or above will produce the disease Blackheart. In October the seed is graded and placed in the chitting houses, the temperature falling gradually from 15·6°C to 4·4°C by early February. Expensive houses, with automatic temperature control, are considered unnecessary, because this range corresponds approximately with the mean atmospheric temperature for the period. All doors are left open during warm spells and heaters are put on during frost. If there is also adequate fluorescent lighting, the sprouts will be short and tough instead of being long, weak, and spindly. The temperature should not be allowed to fall below 2·2°C if possible. A serious set-back in lifting dates will occur following prolonged low temperatures during the chitting stage. Indifference to warm, humid conditions often results in the development of stem canker. Aphids also multiply very rapidly, if present, spreading virus diseases.

During the planting period, weather knowledge is very important, since any mistake made at this stage cannot be corrected later. With the increasing dependence on mechanisation and artificial fertilisers, such mistakes, multiplied many times over, through automation, can prove very costly. Heavy implements should not be taken on to the land until the soil is dry enough or poor results will be obtained. Fertilisers should not be applied too far in advance of planting, in case a period of heavy rain intervenes, with consequent leaching of valuable nutrients into the subsoil. Seed potatoes should not be left in the fields uncovered, if there is a risk of overnight frost. Seed potatoes should not be planted in wet or frosted ground. Extra effort has to be made to take advantage of good planting conditions when the weather is not settled. Sprays, when considered necessary, are more effective, if applied in good growing weather, on calm, cloudy days.

During the growing period, it is inadvisable to irrigate too soon, even under dry conditions, if rain is imminent. Excess moisture in hot weather causes too much haulm growth, delayed maturity, and deterioration in the quality of new potatoes. Forced, quick growth results in a weak cell structure and a low starch content, leading to disintegration and discolouration of the new potatoes when cooked. A foreknowledge of a late spring frost is valuable where it is possible to protect the crop by irrigation or other means. The colossal expense of protecting large areas of potatoes against frost cannot be justified. A careful watch should also be kept on weather reports from other early potato areas during this period. Heavy rain, drought, and late frosts often influence the date of lifting, and consequently market prices, to a marked degree.

It is during the lifting of the potato crop, however, that the greatest benefit can be derived from weather study. Market prices often fluctuate very much from day to day, according to the supply and demand. The weather at the time is an important contributory factor. During wet periods lifting is interrupted, causing 'weather markets' and rising prices. Cold weather increases demand, whereas in hot, sunny periods, consumers prefer salads. It is often better to stop lifting when supplies are plentiful, to avoid ridiculously low returns. Lifting has to be accelerated irrespective of prices if the weather is favourable to the spread of blight. If the spores from diseased plants reach the tubers during rainy periods, the crop might well become unsaleable. Hot weather can easily affect the keeping quality of new potatoes, unless they are disposed of quickly and adequate protection provided. Weather knowledge is also useful in cutting out unnecessary transport costs, both in the conveyance of potato pickers and in long-distance lorries.

To be profitable, early potatoes should be produced at the least possible expense; the crop must be true and uniform with good weight per acre at lifting time; the quality and presentation of the produce should command the best prices the market can provide. It is exceedingly difficult to estimate the full value to the grower of 'competence in weather forecasting', and its application when making decisions, but there is no doubt that it can be very substantial.

Correct Interpretation of Official Weather Forecasts
Short-term forecasts, issued at intervals throughout the day, present no problems. They are precise, reasonably accurate, and well presented. As a result of considerable advance in recent years, they are now generally accepted and used to advantage by people in all walks of life. Monthly forecasts, issued fortnightly, are apparently still in the experimental stage. They are not yet sufficiently precise or accurate to be of any use for decision-making.

The mid-term forecast is what farmers, generally, are mostly concerned with, because plans have often to be made several days in advance. So far, the official mid-term forecast is not sufficiently reliable for making vital decisions. This is where a basic knowledge of meteorology is essential. Daily observations over a long period of years, of cause and effect in the movements and interactions of high and low pressure areas, and their association with particular events in the farming operations at the time, leave an impression on the subconscious mind. This stored knowledge is the source of intuition when assessing the probability of accuracy in official weather forecasts. A satisfactory conclusion and a vital decision should be made only after the weather picture as a whole is viewed in perspective, and a personal assessment made for the particular location concerned.

Present-day Facilities for Acquiring Weather Information
The Meteorological Office issues a detailed weather chart daily, giving pressure distribution and weather conditions in the northern hemisphere. The leading London newspapers also include daily isobaric maps and weather trends. These are most valuable in conjunction with weather forecasts on radio and television throughout the day, covering all land and sea areas around the British Isles. A broad picture is thus obtained of conditions in general. In addition, when in doubt, a local forecast can be obtained by telephone from the nearest Meteorological Office. Last but not least is the importance of possessing a first-class barometer. Sudden changes in atmospheric pressure and deviations in the predicted movements of cyclones and anti-cyclones can be detected in this way. A good barometer will indicate a noticeable fall of pressure when taken up a flight of twenty steps. Observations of wind strength and direction, cloud formations, sun, moon, etc, are useful corroborative evidence, but they must always be viewed as part of the whole weather picture.

Presentation of Weather Forecasts on Radio and Television
The BBC is providing a very useful service in presenting weather forecasts to meet the special needs of agriculture. There are, however, some shortcomings which should be mentioned.

1   The isobars on all television charts should show their barometric value in millibars so that viewers can ascertain the intensity of pressure areas.

2   The late night television chart could be shown earlier, at a fixed time.

3   The afternoon television forecast and chart should be brought forward from 1.45 pm to 12.45 pm, so that it could be viewed during the lunch hour.

4   The shipping forecast on the radio is also ideal for farmers, both in content and time of day, because it gives a very comprehensive picture of pressure distribution and the latest weather conditions over all sea areas surrounding the British Isles. It is therefore most annoying and frustrating that it should currently coincide with a very important farming feature on another channel.

5   The 6 pm television forecast is undoubtedly the most important one of the day for farmers, as it is then that decisions for the following day are usually made. At present, the isobaric chart is shown for a few seconds only. No one can possibly absorb the details in so ridiculously short a time.

Conclusion

It is evident that weather study and forecasting will become increasingly important to British agriculture in the competitive years which lie ahead, particularly when Britain joins the Common Market. The European farmer is often blessed with more sunshine and better weather for cultivating his land and harvesting his crops. A reasonably accurate foreknowledge of the weather conditions necessary to suit the various operations in farming can mean success instead of failure, if costs can thus be reduced by greater efficiency, and returns improved by better quality produce. To assist him in this matter, the farmer has no better friend than the weatherman. It is wise, therefore, not to be too critical when forecasts are incorrect. Anticyclones and depressions can be very elusive, increasing or reducing their rate of travel at short notice. They often change course, deepen, or fill up for no apparent reason, to upset the most carefully prepared computerised forecast. Warm and cold fronts may slow down as they approach and overstay their welcome. An intense depression over Newfoundland may not be destined, after all, to reach our shores, whereas an innocent looking shallow trough west of Ireland may deepen sufficiently to provide enough rain to flood our fields.

In his own interest, therefore, the farmer should endeavour to take full advantage of the facilities now available, to increase his knowledge of this intricate and absorbing subject, which cannot fail to bring him increased prosperity and satisfaction. In so doing, he will also learn to understand and appreciate the complexities of modern weather forecasting.

CHAPTER EIGHT

MARJORY G. ROY
and J. M. PEACOCK

Seasonal Forecasting of the Spring Growth and Flowering of Grass Crops in the British Isles

Introduction

During the spring months in the British Isles the growth of grass and other forage crops is often limited by climatic conditions such as low temperatures at a time when the light available for photosynthesis is quite high. Growth may also be restricted due to waterlogging of the soil, and occasionally it may be affected by soil-water deficit. A further factor which may affect the spring growth is a lack of nitrogen in the soil following winter leaching. In general the nitrifying bacteria in the soil, which are affected by temperature, become active only after grass growth has started, and nitrogenous fertiliser must be applied in order to make maximum use of these periods of potential growth.

If an attempt is to be made to forecast for any year the expected growth of a forage crop in spring, including the increase in harvestable dry matter, and the approximate date of heading (50 per cent ear emergence), it is necessary to define a model for the response of the crop to a given sequence of climatic conditions. In this chapter attention will be concentrated on the perennial grasses, and results of experiments at the Grassland Research Institute at Hurley and the Welsh Plant Breeding Station at Aberystwyth have been used to devise a simple model of the crop growth. Certain routine meteorological observations have been used to estimate the growth in recent years.

The Spring Growth of Forage Grasses

During the winter the grass crop is in the vegetative phase, but in the spring floral initiation takes place, and the reproductive phase occurs with rapid stem elongation of fertile tillers. The time of initiation depends mainly on photoperiod (Cooper, 1952) and occurs in March in many early-flowering varieties but not until April or May in late-flowering varieties. Heading (defined as 50 per cent ear emergence of fertile tillers) occurs in May for cocksfoot (*Dactylis glomerata*) and early-flowering ryegrass (*Lolium perenne*), but not until June for timothy (*Phleum pratense*) and late-flowering ryegrass. Considerable differences in the heading date of a single variety may occur between different sites and in different years, as the rate of floral development depends markedly on the temperature (Cooper, 1952).

The usual definitions of 'earliness' or 'lateness' of the spring growth of different grass varieties are largely related to the relative timing of floral initiation. Visible differences occur in the height of the sward, and the extension of a greater proportion of the plant above cutting height (about 5 cm) leads to a more rapid increase in the yield of dry matter of early varieties as measured by conventional methods of sampling (Davies and Calder, 1969). If March is warm enough for considerable growth to take place, then the apparent differences in the timing of spring growth of early and late varieties may be quite large. However, the increase in dry matter of the total aboveground parts of the plant may be approximately the same in both early and late varieties.

Data from Hurley have shown that at the end of February the leaf-area per unit-land-area (leaf area index) (Watson, 1947a) of an intensively managed perennial ryegrass sward is usually between 1 and 2, and that about 40 per cent of the incoming sun and sky radiation is intercepted by the crop. As the leaf area index increases to about 6 or 7, the efficiency of light interception increases to about 95 per cent. Since light energy is not limiting during this spring period the increase in above-ground material is largely determined by the rate at which the leaves can expand. A fully comprehensive model of crop growth in terms of the increase in dry weight of the various parts of the plant would have to include the calculation of gross photosynthesis and respiration, and the distribution of assimilates between below- and aboveground parts. This has not been attempted in the present analysis. Instead it has been assumed that there is a relationship between the increase in the leaf area index and the increase in fresh weight of the crop. To a certain extent a similar relationship will exist between the leaf area index and the dry weight

(Watson, 1947b). Once rapid elongation of fertile tillers is taking place, the dry weight will also include a considerable contribution from stem material.

In a grass sward the unit of growth is the tiller, and although the grass plant is perennial, the majority of tillers are annual (Langer, 1956; Williams, 1970). The rate of production of dry matter (DM) of a sward is determined by the rate of formation of new tillers, the growth of leaves on existing tillers, the rate and time of flowering, and the death of older tillers (Langer, 1963; Ryle, 1964). A diagrammatic section of a vegetative tiller is shown in Fig 8.1. This shows fully expanded leaves (a), and visible expanding leaves (b).

In the model to be described the figures quoted refer to perennial ryegrass cv. S.24, but in general could be applied to most varieties of temperate perennial ryegrass. The number of live grass tillers per unit area of sward has a distinct annual pattern (Garwood, 1969). Although the numbers of tillers increase rapidly during late winter and early spring in all species, to a first approximation there is about one tiller per cm^2 during the spring period. Ryle (1964) and Peacock (unpublished) showed that the number of actively growing leaves was one per tiller and that the mean width of these leaves was 1·8 mm during the spring period. The increase in leaf area index, L, per day can be written as:

$$\delta L = k(\delta l) - c \qquad (1)$$

where δl is the increase in leaf length in mm day^{-1} and c represents a constant loss of material per day. Using the figures quoted above for perennial ryegrass cv. S.24 a value for k has been calculated.

$$k = \frac{\text{leaf width} \times \text{number of expanding leaves per tiller} \times \text{number of tillers}}{\text{ground area}}$$

considering dimensions and units of length

$$k = 1·8 \text{ mm} \quad \times \quad 1 \quad \times \quad 0·01 \text{ mm}^{-2}$$
$$\text{(text)} \qquad \text{(text)} \qquad \text{(text)}$$

$$= 0·018 \text{ mm}^{-1}$$

Thus

$$L(t) = L(0) + k\Sigma\delta l - ct \qquad (2)$$

where $L(0)$ is the leaf area index at time $t = 0$.

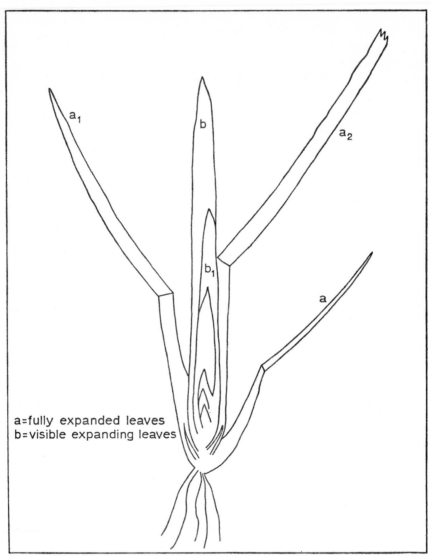

Fig. 8.1   A diagrammatic section of a vegetative tiller

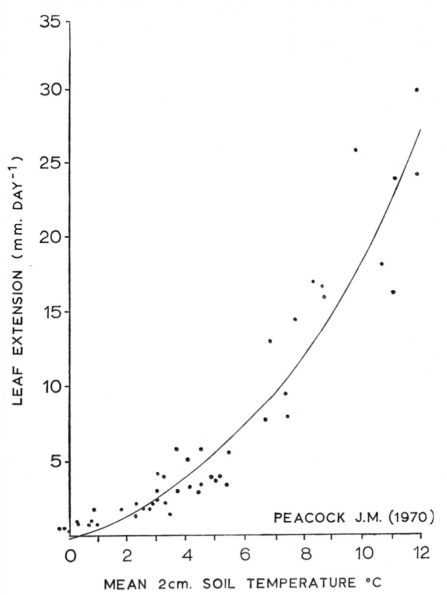

Fig. 8.2 S.24 perennial ryegrass (spring 1970): the relationship between rate of leaf expansion and mean 2 cm soil temperature

Relations Between the Rate of Leaf Extension and Mean Daily Soil Temperatures

Peacock (1971) has shown recently that when soil water and nutrition are adequate, the rate of extension of the youngest leaf on a tiller of perennial ryegrass *Lolium perenne* cv. S.24 in spring depends mainly on the temperature just below the soil surface. Figure 8.2 shows the values obtained during the spring period in 1970, using soil temperatures at 2 cm, averaged over the period of measurement of the leaf extension. The data have been fitted by an asymptotic curve of the form

$$\delta l = A + BC^T \tag{3}$$

T is the mean soil temperature at 2 cm in °C.

$A = -5 \cdot 0 \pm 2 \cdot 2$

$B = 4 \cdot 7 \pm 1 \cdot 7$

$C = 1 \cdot 173 \pm 0 \cdot 031$

It can be seen that some leaf extension occurs even at very low temperatures, but this may not be sufficient to compensate for the loss of plant material by decay. Above about 5°C the rate of leaf extension increases quite rapidly with increasing temperature, but there is no marked 'growth' threshold at this temperature.

It was found that the response curves were almost identical if the mean daily temperatures were measured at 2 cm, 5 cm, 10 cm, or 20 cm depth. This was due to a close correlation between changes in the mean daily temperature at the different depths. In this particular soil, a sandy loam, the time-lag for the diurnal temperature wave was only 3·5 hours between 2 cm and 20 cm depth, while for the annual wave it would be about 3 days. In certain soils, such as those developed on peat, these temperature lags would be much greater, and could amount to 9 days in the annual wave.

Calculation of the Leaf Area Index Using Temperature Measurements

Daily values of mean soil temperature under a sward of S.24 perennial ryegrass were measured at Hurley for 1968, 1969, and 1970, and leaf area indices were measured during the period of spring growth in the last two of these years. Using the relationship between leaf extension and temperature obtained in 1970, daily values of δl and $\Sigma \delta l$, the accumulated leaf extension, were calculated, starting from 1 March. There were six measurements of leaf area index during the period 1 March 1970 to 25 April 1970 and a multiple regression was fitted to the observed and calculated values to obtain values for the constants k and c in eq (1). The resulting regression for leaf area index was:

$$L = L_0 + 0 \cdot 0147 \, \Sigma \, (\delta l) - 0 \cdot 0290 t \pm 0 \cdot 100 \qquad (4)$$

with $k = 0 \cdot 0147 \pm 0 \cdot 0012$ as compared with $k = 0 \cdot 018$, the value postulated for eq (1). The order of magnitude of k was as expected and the difference in values could be due to less than one leaf per tiller expanding at the calculated rate.

Equation (4) has also been applied to the data for 1969 where the initial leaf area at the beginning of March was considerably lower than in 1970.

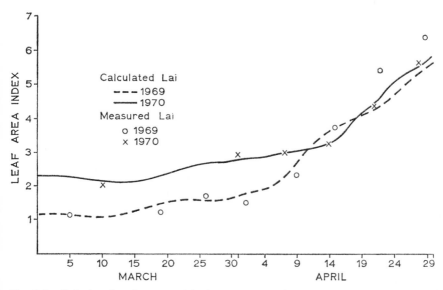

Fig. 8.3 Calculated and measured leaf area indices (High Field, Hurley)

Reasonable agreement was obtained between the observed and calculated values up to 15 April, but on 22 and 29 April the observed leaf area index was considerably higher than the calculations indicated. Figure 8.3 shows the calculated values of leaf area index for 1969 and 1970, together with the actual observations.

Certain deductions can be made using eq (4). If the mean soil temperature is less than 2·4°C (36·3°F) then there is a slow loss of leaf area. The daily increase in leaf area at different temperatures is given in Table 8.1.

Table 8.1

Mean daily soil temperature, °C	Daily increase in leaf area index
3	0·011
4	0·030
5	0·052
6	0·077
7	0·109
8	0·147
9	0·191
10	0·241

Thus an increase in temperature from 5°C to 7°C would double the daily rate of increase of leaf area, and at 10°C it would be almost five times the rate at 5°C.

Dry-matter Yields

In order to obtain an estimate of the above-ground yield of dry matter a regression equation was calculated for 1970 data between the leaf area index and the observed yield of dry matter sampled to ground level. This gave:

$$W = 75 + 80 \cdot 4 L \pm 24 \ \text{g m}^{-2} \tag{5}$$

$$(1 \ \text{g m}^{-2} = 10 \ \text{kg hectare}^{-1})$$

During March and early April most of this material would have been below normal cutting height, and from mid-April about 300 to 350 g m^{-2} would have to be subtracted from the total to give the dry-matter yields sampled by conventional methods. However, as has been pointed out by Davies and Calder (1969), this part of the above-ground yield could be utilised to some extent by sheep which graze close to the ground.

Given the initial leaf area index of a sward of S.24 perennial ryegrass at the beginning of March, and daily values of the mean soil temperature at 2 cm or 10 cm under the sward, it appears that a reasonable estimate can be made of the changes which occur in the leaf area index up to the point where further leaf emergence ceases and the stems of fertile tillers elongate rapidly. (This assumes that nitrogen fertilisation is adequate for maximum growth.) Using the calculated values of leaf area index, an estimate can then be made of the above-ground yield of dry matter. Further growth of the crop depends mainly on the progress of reproductive development and on the incoming light energy, and usually occurs at a fairly constant rate until after the time of ear emergence.

From a forecasting point of view the main information required concerns the changes in soil temperature to be expected during the coming season. Additionally it may be possible, using observed temperatures during February and March to assess the growth that has already taken place (perhaps below normal sampling height) and to use this as a basis for forecasting crop growth during April.

Relations Between Microclimate Measurements Under a Grass Crop and Standard Meteorological Observations

In general, mean soil temperatures are not measured under a grass sward, and it is found that under bare soil, temperatures tend to be lower during cold periods and higher in warm periods due to the lack of an insulating plant cover. In particular the 0900 GMT soil temperature under bare soil is a relatively poor indicator of the mean soil temperature at the same level under a sward. As the leaf area index increases, differences become more marked and can amount to several centigrade degrees. The once-daily observation of temperature at 30 cm under short turf appears to be much more useful, particularly in soils with a large damping depth. The diurnal variation is very much smaller than at 10 cm under bare soil.

At Hurley, readings were made daily in 1970 at 0900 GMT of the maximum and minimum temperatures at 10 cm depth under turf, using a mercury-insteel thermometer with the bulb lying horizontally at the required depth. A linear regression equation was calculated between daily values of $\frac{1}{2}$(minimum +maximum) temperature as *read* at 0900 GMT and the 0900 GMT 30 cm earth temperature under turf. The months used were March, April, and May. The equation obtained was:

$$T_{\text{mean}} = 0 \cdot 24 + 1 \cdot 12\, T_{30} \pm 0 \cdot 62°C \qquad (6)$$

T_{mean} = mean soil temperature at 10 cm

T_{30} = 0900 GMT earth temperature at 30 cm

Changes in the diurnal range of temperature at 30 cm, as the mean temperature increases, and the lag of the annual temperature wave, affect the constants obtained, and these would be rather different for soils with a smaller damping depth and larger phase differences.

These observations were made in the meteorological enclosure at Hurley, about 40 m lower than the field where leaf extension and temperature have been measured in the last two years. Although the height difference is quite small, differences in soil temperature during the springs of 1969 and 1970 have been between 1°C and 2°C, probably due to the greater exposure to wind of the higher site. As a result there is likely to be a delay of between 7 and 14 days in the date at which a given leaf area index is reached at the higher site as compared with the lower one.

Calculated Leaf Area Indices 1964–70
Figure 8.4 shows the values of leaf area index at 5-day intervals during March and April for the years 1964 to 1970, calculated from eq (4). The initial value on 1 March was taken as 1·5 for each year. The mean soil temperatures were obtained from daily values of the 30 cm earth temperature in the meteorological enclosure using eq (6). These data show that only in 1966 and 1967 would considerable growth have taken place during March, and the decade 1961 to 1970 has been notable for the number of years (seven) in which little grass growth would have occurred before the end of March in most of England and Wales. In 1965, the late start to active growth was compensated to some extent by a warm spell of weather in late March and April.

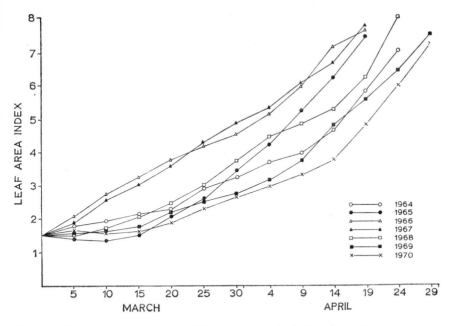

Fig. 8.4 Calculated spring growth (Hurley meteorological enclosure)

In order to test how accurate a mean monthly earth temperature would be as an indicator of the increase in leaf area during March, the mean monthly earth temperatures at 0900 GMT were converted into mean monthly soil temperatures at 10 cm using eq (6), and a mean value of daily leaf extension was calculated. The resulting changes in leaf area between 1 and 31 March, using daily values and mean monthly values for δl, are given in Table 8.2.

The use of the mean monthly temperature gives an under-estimate of the increase in leaf area in months in which there is a large rise in temperature, such as March 1965, when the minimum value during the month was 1·0°C and the maximum 8·9°C. This is because the rate of leaf extension is a non-linear function of the mean soil temperature. However, in general the mean monthly earth temperature under grass is a useful indicator of the growth to be expected during the month, with the proviso that it is likely to under-estimate the final total under certain conditions.

Table 8.2

	Change in leaf area using mean monthly temperatures	Change in leaf area using mean daily temperatures
1964	1·74	1·95
1965	1·56	2·08
1966	3·20	3·20
1967	3·48	3·47
1968	2·24	2·42
1969	1·29	1·35
1970	1·10	1·22

Forecasting the Mean Monthly Earth Temperature

From February to March there appears to be a fair degree of persistence in the 30 cm earth temperatures, and a multiple regression has been calculated using the observed 30 cm earth temperature for February, and the mean monthly air temperature for March. At Hurley this was:

$$T'_{\text{earth}} = 0·21 \ T_{\text{earth}} + 0·64 \ T'_{\text{air}} + 0·72 \pm 0·54 \tag{7}$$

where T refers to mean February temperatures and T' to mean March temperatures. Thus a forecast monthly mean air temperature for March can be combined with the observed February earth temperature to provide a forecast of the March mean earth temperature. Because of different site and soil effects this regression should be calculated, where possible, for the site where it is to be used.

Heading of Forage Grasses

The time of heading (the date when 50 per cent of the fertile tillers have reached ear emergence) of forage grasses is of interest as an indicator of changes in the digestibility of the crop. The relationship between digestibility and the stage of reproductive development varies with species and also with varieties within a given species (Aldrich and Dent, 1967; Green, Corrall, and Terry, 1971). For example the digestibility of a cocksfoot is generally lower than that of a perennial ryegrass which heads at the same time; and within the ryegrasses the digestibility at the time of heading is higher in early-flowering varieties than in the later ones.

Cooper (1952) showed that much of year-to-year and site-to-site differences in heading dates could be explained by a consideration of the air temperatures between the time of floral initiation and ear emergence. He used accumulated day-degrees above 42°F (5·6°C) to represent the temperature effect and found that ear emergence of spaced plants of S.24 perennial ryegrass could be related to temperature by a regression of the form

$$Y = 30 \cdot 23 - 42 \cdot 4 \, \log_{10} \left(\frac{T}{100} \right) \tag{8}$$

where Y was a date in May, and T was the accumulated temperature above 42°F (5·6°C) during March and April. (1 Fahrenheit day-degree is equivalent to 0·555 Celsius day-degree.) A similar regression was obtained using accumulated day-degrees for April and May to give the date of heading in June of S.23 perennial ryegrass.

These regressions have been tested with some recent data of 50 per cent ear emergence recorded by the National Institute for Agricultural Botany at their testing stations at Seale Hayne (Devon), Trawscoed (Cardiganshire), Cambridge, and Cockle Park (Northumberland); and by the Scottish Agricultural Colleges at Ayr, Edinburgh, and Aberdeen. Quite reasonable agreement was obtained between the calculated and observed dates for S.23 perennial ryegrass, but with S.24 considerable differences occurred, particularly at the more northerly centres in 1967 and 1968 where very low temperatures during May led to late heading dates.

The influence of latitude has sometimes been confused with that of climate, but the latitude in itself will mainly affect the changes in daylength. Unless floral initiation occurs before the middle of March the appropriate daylength will be attained at an earlier date at more northerly latitudes, and by May such differences will be considerable. Floral development is accelerated by longer daylengths (Aitken, 1966). In years which have warm springs throughout the British Isles (such as 1957 and 1961) differences between the heading dates at northern and southern sites are small, although the accumulated day-degree total may be somewhat smaller in the north. The effects of altitude and exposure over a height range of 300 m may be as great as the average differences at sea-level between the south-west and the north-west of the British Isles. More useful results can be gained by concentrating attention on long-term climatic data for a given site together with expected deviations from these, rather than by looking for a general south-to-north variation.

The range of heading dates of S.24, which may be taken as being typical of the early-heading perennial grasses, in the nine-year period from 1962 to 1970 has been about 14 days at Cambridge and Seale-Hayne and 26 days at Aberdeen, and this period does not contain any spring that would be classified as 'early' as far as heading is concerned. In 1966 and 1967 mild weather in early March was followed by some quite cold spells in late March, April, and May. One result of a late start to spring growth followed by a warm or average late April and May is that the dry-matter yields may be a fortnight or more behind those obtained in an 'average' year throughout the growth period, but the later stages of floral development are accelerated by the high temperatures. As a result the delay in heading is considerably smaller than that in reaching a given yield and there is a corresponding loss in the yield of digestible dry matter. A forecast of the probable date of ear emergence could, therefore, be of practical interest.

Following on from the results obtained with respect to leaf extension, it would appear that during the early stages of floral development, after initiation has taken place, the important temperature would not be that of the air at Stevenson Screen level but that of the soil just below the surface. Similarly, the form of response to temperature is likely to be an asymptotic one. Once rapid elongation is taking place and the developing inflorescence is carried up to near the top of the crop, then air temperatures may provide a reasonable estimate of the temperature affecting the inflorescence. The importance of soil temperature during the early stages may explain why the total of accumulated day-degrees up to heading date tends to be lower near the west coast than it is in the east. In the former areas as compared with the latter, mean soil temperatures are generally higher during March, but mean air temperatures tend to be lower during April and May.

In assessing the effect of temperature on the date of heading and hence devising some scheme of forecasting, a single agronomic observation is being compared with a sequence of climatic conditions which may be of varying importance at different stages in the development of the inflorescence. Additionally, the weather during the preceding winter and the management of the sward may affect the heading date. A calculation of accumulated day-degrees above 5·6°C will give an indication at the end of March as to whether the heading dates of early varieties may be earlier or later than average, but probably the value of $\Sigma(\delta l)$, which uses an asymptotic relationship with mean soil temperatures, would be a better predictor at this stage. It may be useful to estimate the date at which $\Sigma(\delta l)$ reaches a value of about 350 (corresponding to a leaf area index of around 5·5), but it must be remembered that changes in air temperature at crop level after this time could have a significant effect on the time of heading.

Forecasting Requirements for Estimating the Spring Growth and Date of Heading of Perennial Grasses

1 Mean soil temperature during March under a grass sward, probably from a forecast of the mean earth temperature at 30 cm depth under short turf. If the month is expected to be very cold at the start then some estimate is required of when the temperature is expected to rise substantially (if at all).

2 Mean air temperature during April; also mean soil temperature during April if March temperatures are expected to be cold at the site in question, and if the grass concerned is of a late-flowering variety.

3 Mean air temperature during May.

Considerable information on the progress so far during a given season could also be calculated from actual observations of temperature, and by the end of March it should be possible to make some estimate of whether the heading of early varieties is likely to be early or late.

References

AITKEN, Y. (1966). The flowering response of crop and pasture species in Australia. 1. Factors affecting development in the field of the *Lolium* species. *Aust. J. Agric. Res.*, **17**, 821–39.

ALDRICH, D. T. A. and DENT, J. W. (1967). The relationship between yield and digestibility in the primary growth of nine grass varieties. *J. Natn. Inst. Agric. Bot.*, **11**, 104–13.

COOPER, J. P. (1952). Studies on growth and development in *Lolium*. 3. Influence of season and latitude on ear emergence. *J. Ecol.*, **40**, 352–79.

DAVIES, A. and CALDER, D. M. (1969). Patterns of spring growth in swards of different grass varieties. *J. Br. Grassld Soc.*, **13**, 215–25.

GARWOOD, E. A. (1969). Seasonal tiller population of grass and grass/clover swards with and without irrigation. *J. Br. Grassld Soc.*, **24**, 333–44.

GREEN, J. O., CORRALL, A. J., and TERRY, R. A. (1971). Grass species and varieties. Relationship between stage of growth, yield and forage quality. *Tech. Rep. 8 Grassld Res. Inst. Hurley.*

LANGER, R. H. M. (1956). Growth and nutrition of timothy (*Phleum pratense*). 1. The life history of individual tillers. *Ann. Appl. Biol.*, **44**, 166–87.

LANGER, R. H. M. (1963). Tillering in herbage grasses. *Herb. Abstr.*, **33**, 141–8.

PEACOCK, J. M. (1970). Plant and crop growth and the environment. Interaction between the sward and the environment in the field. *A. Rep. Grassld Res. Inst.* 1969, p 63.

PEACOCK, J. M. (1971). Plant and crop growth and the environment. Interaction between a crop and the environment in the field. *A. Rep. Grassld Res. Inst.* 1970, pp 55–7.

RYLE, G. J. A. (1964). A comparison of leaf and tiller growth in seven perennial grasses as influenced by nitrogen and temperature. *J. Br. Grassld Soc.*, **19**, 281–90.

WATSON, D. J. (1947a). Comparative physiological studies on the growth of field crops. 1. Variation in net assimilation rate and leaf area between species and varieties and within and between years. *Ann. Bot. N.S.*, **11**, 41–76.

WATSON, D. J. (1947b). Comparative physiological studies on the growth of field crops. II. The effect of varying nutrient supply on the net assimilation rate and leaf area. *Ann. Bot. N.S.*, **11**, 375–407.

WILLIAMS, R. D. (1970). Tillering in grasses cut for conservation, with special reference to perennial ryegrass. *Herb. Abstr.*, **40**, (4), 383–6.

CHAPTER NINE   R. A. BUCHANAN

Weather Forecasting for Industry

In the main the first part of this book has dealt with the use of weather information in agricultural activities. Now we turn to consider applications
of weather knowledge in some other human activities which the title of the
book suggests can be classed under the heading 'industry'. A glance at the
titles of some of the chapters to follow will show how wide the field is:
tourism, recreation, road construction, weather hazards to traffic, beer
drinking—a wide and varied field indeed and indicative of the pervasive
influence of weather.

This chapter will be concerned chiefly with the construction, manufacturing,
and service industries in the United Kingdom and with the forecasting
services that they require. But before dealing with these particular requirements it would perhaps be profitable to outline the forecasting services that
are provided in the United Kingdom to meet the general public interest in
weather.

General Services (See Appendix p 232)
It is the duty of the Meteorological Office, as the state weather service, to
keep the community informed of current and expected weather in and around
the United Kingdom. This general function is discharged by providing
weather bulletins for transmission by radio and television and for publication
in newspapers. The time devoted to weather on radio and television is quite
considerable. All four BBC radio channels broadcast weather forecasts
prepared by the Meteorological Office. On these main channels the total
transmission time devoted to weather, excluding special forecasts for shipping,
is about four and a half hours each week. On BBC television, the weatherman
appears three times daily on BBC 1 using isobaric charts and maps with
legends to describe the weather sequence. At other times, both on BBC 1 and

115

BBC 2, scripted forecasts are read while caption charts are shown on vision. Most of the independent television companies show weather forecasts in their programmes.

From figures provided by the BBC Audience Research Department estimates may be obtained of the number of people who listen to or watch the weather forecasts. In 1970 the peak radio listening time on weekdays was in the morning, around breakfast time, when about ten million people heard the weather forecasts. At weekends the pattern was different—audiences were smaller and on Sundays the mid-evening weather bulletins were the most favoured. On BBC television, viewers of the weather presentation also averaged around ten million each night. Figures for independent television are not available but it seems reasonable to assume that they are of the same order as those for BBC television. Allowing for some overlap of listeners and viewers it emerges that 20 million or more people—40 per cent of the population or over—keep in touch with the weather developments daily through these two mass communication channels.

Most of the weather bulletins sent out by these means are, inevitably, general in character and limited in their detail by the time available on the channel. With some exceptions they can provide only broad descriptions of the weather, usually for fairly large areas of the country. Nevertheless they do ensure that anyone with an interest in the weather can keep himself informed about its broad features, whether the interest arises out of curiosity only or because some decision may be influenced by weather considerations.

There are some important exceptions to the general rule that weather information on radio and television can be given only in broad terms, and the exceptions illustrate the value of these channels as a ready and quick way of advising the public of adverse weather likely to affect large sections of the population. With the co-operation of the BBC and the ITV companies, warnings of the occurrence of dangerous conditions such as dense fog, moderate or heavy snow, and glazed frost or icy roads are issued on Radio 2, BBC TV and most ITV channels. These are known as FLASH messages. In addition forecasts of fog or strong winds affecting motorways are broadcast on Radio 2 as occasion demands and, for the benefit of local authorities with road clearance responsibilities, warnings of expected snow are transmitted on Radio 2 at 4 pm.

It should be emphasised that, with the exceptions just mentioned, the weather bulletins issued on the mass communication channels are general in character and usually for fairly large areas of the country. However, many activities are very sensitive to weather conditions, and, for these, detailed advance information about the weather can often assist in contingency planning to allow advantage to be taken of favourable weather or to avoid some of the consequences of adverse weather. Two sources of weather advice are available to serve these special interests.

I. The Automatic Telephone Weather Service
This service, operated by the Post Office, is available from about fifty telephone centres in the United Kingdom. At each centre tape-recorded forecasts for the local area are maintained. The forecasts are provided by the Meteorological Office and are changed several times daily, keeping them up to date. These arrangements mean that anyone with access to a telephone and living in one of the areas served, can obtain an up-to-date forecast for the local area at any time of the day or night. Over thirty million people live in the areas covered by the service, which the Post Office is still extending. In 1969 some $12\frac{1}{2}$ million calls were recorded on the service.

II. Special Personal Services
In the services described so far—radio, television, automatic telephone weather service—there is no personal contact between the customer and the meteorologist. The general type forecasts on radio and television, for fairly large areas of the country and without much detail of timing of weather events, may satisfy the needs of many of the listeners and viewers. The rather more detailed forecasts on the automatic telephone weather service may cater for some whose activities demand a closer knowledge of the likely progress of the weather. But there remains a substantial number of people whose needs are not satisfied by these impersonal services and who seek the personal advice of a forecaster on their problems.

To cater for this desire for special personal service the Meteorological Office offers facilities through about forty of its stations in the United Kingdom. The telephone numbers of these stations are given in area telephone directories and are publicised in other ways. In addition some fifty other stations have contacts with the public and deal with weather inquiries. The demand for personal service provided by these offices has increased year by year. By the late 1950s it became clear that in some areas of the country this growing volume of work could not be handled satisfactorily by offices whose main function was weather service for aviation. As a result special offices were opened in large cities with the aim of meeting the needs of the community for specialist advice not provided on the public information channels. The first office of this type was the London Weather Centre, opened in 1959 on a ground floor site in Kingsway with a shop window and ready access by the public. This office subsequently moved to High Holborn. In December 1959 a Weather Centre was opened in Glasgow, followed in June 1960 by one in Manchester and in December 1961 by one in Southampton. In April 1967 the forecasting office at Watnall, on the outskirts of Nottingham, which had previously served aviation, became a public service office with functions similar to those of a Weather Centre. In the same month the Newcastle Weather Centre was opened.

From records maintained by offices it is possible to study the growth of inquiries over the last nine years, the period for which comparable figures are available. In 1962 the total number of inquiries handled by all stations of the Meteorological Office was just under 800,000. Each succeeding year saw an increase and in 1970 the total reached about 1,650,000, more than double the 1962 figure. Of these grand totals the share of Weather Centres increased from between 45 and 48 per cent in the earlier years to 55 per cent at the end of the period. Very nearly all of these inquiries concern short-term weather prospects, ie, for periods up to one or two days ahead.

A monthly breakdown of the annual totals shows seasonal peaks each year in mid-winter and in the summer. There is some variation from year to year in the particular months in which the highest and lowest numbers occur. Variations of this kind can be traced to unusual weather, perhaps the most notable case being the effect of the snowstorms of February 1969 on the demand for weather advice, which produced 203,820 inquiries, the highest monthly total yet recorded. At the beginning of the period 1962 to 1970 the lowest monthly totals, in the spring and autumn minima, numbered about 50,000, the highest monthly totals, in winter and summer, about 100,000. By the end of the period the 'quiet' months were producing totals of 100,000 a month and the busy ones totals of 170,000, with on one occasion over 200,000 as already noted.

Users' Interests

In the recording system used at offices providing a personal service, each inquiry is allocated to an appropriate category describing the user's interest, as far as this is known. These categories can be grouped into five main classes: agriculture; industry; holidays and recreation; marine; road transport.

Table 9.1 gives the total number of inquiries in each of these five classes in the years 1962 and 1970, and the 1970 total expressed as a percentage of the 1962 total.

Table 9.1

NUMBER OF INQUIRIES, BY CLASSES, RECEIVED BY THE METEOROLOGICAL OFFICE IN 1962 AND 1970

	Agriculture	Industry	Holidays & Recreation	Marine	Road Transport
1962	82 826	134 874	114 820	57 948	107 907
1970	148 048	323 073	329 947	141 840	190 928
$\frac{1970}{1962}$%	179	240	287	245	177

In all five classes there has been a substantial increase. Intervening years show the same general tendency but with fluctuations in the rate of increase from year to year. These fluctuations differ from class to class, the more erratic ones being agriculture and road transport, both of which react noticeably to periods of unusual weather. By contrast, the other three classes have maintained fairly steady rates of increase, perhaps surprisingly so in the case of holiday inquiries which one might have expected to vary at least to the same degree as agriculture and road transport.

Industrial Inquiries

The industrial class in Table 9.1 consists of three categories—the construction industry, manufacturing industries, and public utilities. This last category includes the gas and electricity undertakings, and local authorities. Table 9.2 gives the number of inquiries in these three categories in the years 1962 and 1970 and the 1970 totals as a percentage of the 1962 totals.

Table 9.2

NUMBER OF INDUSTRIAL INQUIRIES, BY CATEGORIES, RECEIVED BY THE
METEOROLOGICAL OFFICE IN 1962 AND 1970

	Construction	Manufacturing	Public Utilities	Total
1962	28 292	33 714	72 868	134 874
1970	97 111	82 279	143 683	323 073
$\frac{1970}{1962}\%$	343	244	197	240

The Construction Industry

The most notable feature in Table 9.2 is the much larger increase in the 'construction' category than in the other two categories. In 1962 'construction' inquiries accounted for 21 per cent of the industrial total. In 1970 the 'construction' share was 30 per cent of a total which had itself risen to 2·4 times the 1962 value. Most of the inquiries arise during the construction stage of building and civil engineering contracts, when advance knowledge of the weather can help in the day-to-day planning of the work, in organising the timely delivery of materials and also by warning of the approach of adverse weather which may interrupt work, damage newly poured concrete and stored materials, or give rise to dangerous working conditions on the site.

To assist the site manager in facing these problems the Meteorological Office offers a choice of two types of detailed service. First, there is a warning service under which the contractor registers with the appropriate forecasting office for the area and is then advised whenever the weather he specified is expected to occur in the near future, such as moderate or heavy rain, snow, frost, strong winds. Alternatively, arrangements can be made for a routine supply of forecasts for the particular area, once daily or more often, and these will include any weather elements of special interest at the different stages of construction, for example frost risks at the concrete-pouring stage and the likelihood of strong winds when tower cranes are in use on the site.

For large construction sites a comprehensive service is available that includes both of these types of forecast. Twice daily, at agreed times, there is a discussion between the site manager and the forecaster on the weather expected in the next twenty-four hours or so and the outlook beyond. These personal talks give an opportunity for the two to talk over the aspects of the weather of importance on the site at the time. If sudden or unexpected developments in the weather take place the forecaster telephones the site at once and discusses them. By this personal contact and also by visits to the site the forecaster follows the progress of the contract and also gets to know any weather peculiarities of the site. To a large extent he becomes a member of the construction team as the adviser on weather.

This comprehensive weather service has been used for a wide variety of construction contracts—large buildings in cities, housing estates, a dam in the Pennines with its approach roads, motorway construction, the laying of a new sewer system in a large town in the Midlands, the laying of a trunk gas main in East Anglia for the distribution of North Sea gas.

Manufacturing Industries

In this category the main problem in studying the demand for weather forecasts, and trying to satisfy it, is the very wide range of activities covered. Indeed it would be difficult to name a manufacturing industry that was not represented among the customers for weather advice.

The requests are prompted by two different considerations. In the first the manufacturing process is itself affected by weather. Many processes work best, or may only work at all, if the temperature or humidity (or both) lie within particular ranges. A few examples will show the diversity of activities affected. In textile manufacturing, man-made fibres are less tolerant of fluctuations in temperature and humidity than natural fibres when being woven. In bread-making there are optimum ranges of temperature and humidity for both the leavening and proving processes and these ranges are higher than those in which the flour and yeast should be stored before use. The manufacture of rubber-backed carpets is carried out under carefully controlled conditions of heat and humidity. Of course the processes quoted all take place under cover and the problems for the manufacturer are often air-conditioning ones. But the efficiency of air-conditioning can be aided by a knowledge of the expected variations in the free atmosphere.

The second consideration that leads manufacturers to seek weather advice is in the marketing field. The sales of many products are sensitive to weather and an understanding of how the public's buying habits react to the weather, plus a forewarning of the weather to come, can help a manufacturer to have the right product in the right place at the right time. This skilful application of weather forecasting is limited to those industries in which the manufacture, distribution, and selling processes can take place within the time beyond which the reliability of the forecast falls below the break-even level of economic success. This is a severe limitation but there are some cases where the technique has been applied successfully, particularly in the catering industry. A number of restaurant managers plan their menus in the light of the expected weather and the dishes their customers are likely to ask for in consequence. An enterprising restaurateur who owns both a chain of fish restaurants and a chain of ice-cream shops varies his wholesale buying and deploys his staff between these two activities, using the weather forecast as a predictor of where the greater demand will fall. His strategy cannot be unsuccessful for he has been a regular customer of the Meteorological Office for some years.

Public Utilities and Local Authorities
The most exacting customers for weather forecasts among the public utilities are the gas and electricity undertakings. This is a reflection of the acute sensitivity of consumer demand to variations in the weather and the consequent need for the supplier to anticipate large fluctuations in demand. The requirements of local authorities for weather advice present problems somewhat different from those set by the gas and electricity industries. In all three

cases, however, efficient weather services, designed to fit the particular needs of each, can contribute greatly to the safety and well-being of the community.

In the *gas* industry the much increased use of gas for domestic purposes in recent years has made the industry more weather-sensitive than ever. Formerly, sudden increases in demand were met by increasing the production of town gas or by drawing on reserves stored in conventional gas-holders. With the introduction of natural gas the problems of meeting gas demand have remained, although they are somewhat different. Natural gas enters the country at three points from which a grid takes supplies to many areas. The gas takes an appreciable time to move across country through the grid and, although the pipe-line system can be used for storage purposes by 'line-packing' within certain pressure limits, changes in gas demand have to be met by increasing or decreasing rates of supply from the North Sea Gas Producers. The rate of extraction from the gas fields can be modified only on several hours' notice from the Gas Council and thus the importance of accurate weather forecasts, needed in order to assess gas demand and to give the required notices, still remains. For these reasons it is important that the industry should have access to the most up-to-date weather information, and the Meteorological Office, in conjunction with the Gas Council and the Area Gas Boards, has established a system of detailed temperature forecasts to meet these needs.

The demand for *electricity* is particularly sensitive to weather. Superimposed on the marked regular pattern of diurnal, weekly, and seasonal variations of electricity consumption there are large fluctuations attributable to weather factors. On a typical winter weekday a general fall in temperature of 1°C increases electricity demand by about 1·9 per cent, or 650 megawatts, nationally. A change in daylight illumination from a clear to a completely obscured sky can send up the demand by as much as 5 per cent at certain times of the year. Changes in wind speed and the onset or cessation of precipitation also have identifiable effects on demand.

The total amount of electricity generating plant is designed to meet the highest winter demand but very large savings in fuel and money can be made by running just sufficient plant at any one time to meet the consumer demand and by maintaining a calculated amount of spinning reserve against expected fluctuations caused by weather changes. Accurate weather forecasts clearly have a vital part to play in the efficient operation of such a system and because of this a very detailed service of weather forecasts is provided to National and Area Grid Control Centres.

A major problem facing *local authorities* with road maintenance responsibilities is the need to keep traffic flowing as smoothly as possible under all weather conditions. For many years the Meteorological Office has operated a scheme during the winter months whereby subscribers are notified when snow or low temperatures are likely to affect road conditions. Year by year more authorities have taken this warning service and in 1970 the total exceeded 400.

Fog is another serious hazard to road traffic, particularly to the fast-moving traffic on motorways. Following multiple crashes on motorways in fog in 1965 a close liaison was established between forecasting offices and county police authorities for the exchange of warnings and motorway weather reports. This scheme has been extended as the motorway network has grown.

Conclusion

A recurring feature of this review of forecasting services provided by the Meteorological Office has been the growing demand over the years. The figures quoted of industrial inquiries for personal weather advice show clearly that industrialists are trying to make more use of the forecaster's knowledge to improve the efficiency of their business by taking weather effects into account. On the one hand we have industry calling for specialist advice on weather and on the other hand the meteorologist, increasingly aware that he has advice to offer that could be economically valuable. How can the demand and the supply be brought together to obtain the greatest benefits?

Too many industrialists know too little, both about the effects of weather on their business and about the range, nature, and limitations of the advice that the forecaster can give. The forecasters in turn are not always as helpful as they might be because they too lack sufficient knowledge of the industrialists' problems to frame their advice in the most helpful way. Because of this lack of communication and of understanding the possible economic benefits are not being fully realised. The solution is clear and each side has a part to play. The industrialist must identify the effects of weather on his business as precisely as possible and for this first step he should seek the help of a meteorologist. Once this step has been taken, both have a closer knowledge of the problem and can seek the best way of meeting it. Only by working together as a team can the industrialist and the meteorologist learn enough about each other's problems to be able to realise the economic benefits that can come from the successful application of weather knowledge to weather-sensitive industries.

F

The Weather Forecaster and the Tourist —
The Example of the Scottish Skiing Industry

Introduction

Discussion of the economic importance of recent climatic changes over the British Isles has mainly been concerned with investigating the effect that the general lowering of temperature and increased variability of rainfall (Lamb, 1969) is having on agriculture. Economists have made some important contributions towards identifying the economic impacts of weather variations, following the specification of the dimension of such changes by the climatologist, but there remains, particularly in the tertiary sector of the economy, a lack of awareness of the economic consequences of climatic changes. It is only by taking into account the points at which weather and climate impinge on the performance of an enterprise that the forecaster can assess the type of forecast presentation that is likely to be of value.

The weather sensitivity of tourism is obvious, and the forecaster has found that in addition to the dissemination of general information, as in the 'Holiday Weather' programmes broadcast in the early morning by BBC radio during the summer season, certain activities require more specific information. Skiing in the Cairngorms, one of the major growth points of the Scottish tourist industry in recent years, is an example of a leisure pursuit of this kind. The three leading ski-resorts of Cairngorm, Glencoe, and Glenshee report snow conditions once per day during the season to the Glasgow Weather Centre, giving details of the state of the runs, the nature of the snow surface and the state of the access roads. This information is transmitted using a ten-point code and includes details for the main runs as well as the lower slopes. Estimates of the maximum runs are also given.

It is on the basis of these reports and the forecasters' knowledge of the developing synoptic situation that a forecast which includes the expected

126

height of the freezing level is issued. It is very difficult to measure the effect thas forecasts have on the flow of visitors, partly because at Cairngorm the tourist infra-structure has been designed to provide counter-attractions when conditions for skiing are not very suitable. The distance of the snow resorts even from the main conurbation of central Scotland does mean that for many potential visitors, trips must be planned, rather than be casual and spontaneous. Clearly, the forecast issued before a weekend or holiday period can be important in encouraging or dissuading visitors.

The Scottish Skiing Industry

The Scottish Ski Club was founded in 1907 and an attempt was made to organise winter sports in Scotland in the 1930s but it failed because of poor snow conditions. Interest in the Cairngorms as a major skiing area began during World War II, when the area was used for military training. It was due to the efforts of local hoteliers and the Scottish Council for Physical Recreation that the development of the Glenmore ski-road in 1961 and the completion of the first section of chairlift in 1962 took place. Since that time growth has been dramatic, including the provision of further hotels, chairlift extensions, and a tourist complex at Aviemore, and it is estimated that 25,000 people now ski in Scotland during the season. Naturally the utilisation and exploitation of natural resources of scenery and snow has been encouraged by the concomitant growth of leisure time, affluence, and mobility but it is also worth remembering that the developments have taken place within the context of the recent climatic fluctuations.

The ski-slopes rely on long-lasting snow beds that accumulate in the immature north-facing corries above 762 m (Green, 1968). Snow cover here bears little relation to snow conditions outside the immediate confines of the corries or at the ski-resorts themselves which all lie at considerably lower altitudes. It is often the case that there is no snow at all at Aviemore and the other resorts in the Spey Valley and only a faint general powdering over the highest peaks, but nevertheless skiers are able to enjoy good sport in the corries. For this reason the length of potential season varies little from year to year, normally commencing in mid-December and ending towards the end of April, although in favourable springs, eg 1967 (Spink, 1968), skiing continues on atrophied snowbeds into mid- or late May.

Weather conditions do affect the length and condition of the main ski pistes, which have been prepared near the chairlifts and tows, and the operation of the lifts and tows themselves. Snow conditions vary according to the weather, being generally wet when warm air from the west arrives, but hard and crisp and at times very icy near the summits, in northerly airstreams. Skiers seem to differ as to the type of surface preferred but in general high winds accompanied by raw damp weather are likely to prove inclement to all but the really dedicated sportsman, while blizzard conditions can occasionally halt skiing altogether, as on 4 February 1968, although the swimming pool and ice rink at the Aviemore centre often do peak business if such weather occurs at weekends (Upper Speyside Report, 1970). Lack of snow can also affect the use of the lower nursery slopes where instruction is given.

Regulations lay down that chairlifts have to close when the cross-wind velocity exceeds 53·1 km/h and ski-tows must cease if the wind force is 98·2 km/h or more. The upper part of the Cairngorm lift is exposed to the south-west and in March 1967 an anemograph there recorded a wind gust of 233·4 km/h. Experience from the last few years suggests that it has to be closed for about 35 per cent of the total time, while the White Lady ski-tow which runs parallel to the chairlift has a winter operating efficiency of approximately 85 per cent. The Glencoe lift, which is well sited, loses far fewer operating days because of high winds (about eight occasions in the last three winters). In general then, a strong airflow from points between south-west and north-west is unhelpful to the skiing fraternity, hindering transportation as well as making conditions unpleasant, with frequent hill fog and low sunshine hours. To be balanced against this is the fact that wind shadow effects in the corries are least when the direction of airflow is between south and west and Spink (1970) has reported that the ski-lift managers had complained that during the winter of 1968/9 persistent airflow from between east and south had given meagre winter snowfalls. Sometimes, eg, in 1966/7, unfavourable wind direction in winter has been followed by springs with much snow, and with the corries offering almost unbroken 'runs' even in June (Spink, 1968). The 1970/1 winter illustrates well the vagaries that can affect skiing—after record high temperatures in mid-January, grass ski equipment was sent to Aviemore, and then in April gales with gusts over 136·8 km/h

Table 10.1

PERCENTAGE FREQUENCY CHANGE OF OCCURRENCE OF
WEATHER-TYPE GROUPS FROM 1910–30 TO 1948–68

Type	Jan	Feb	March	April
Westerly	− 15·8	− 7·6	− 6·6	− 1·6
Cyclonic	+ 2·0	0	− 3·8	− 3·6
Anticyclonic	+ 7·8	0	+ 2·3	+ 5·5
North-westerly	+ 2·2	+ 2·7	+ 1·8	− 0·2
Easterly	+ 2·9	+ 5.2	− 0·8	− 0·9
North-easterly	+ 0·4	+ 1·7	+ 0·2	− 1·0
South-easterly	− 0·6	− 1·4	+ 5·4	− 0·4
Northerly	+ 4·0	+ 5·5	− 1·5	− 0·9
South-westerly	− 3·0	+ 0·1	− 1·3	+ 1·2
Southerly	− 2·2	− 4·9	+ 3·9	+ 1·7
Unclassifiable	+ 2·3	− 1·3	+ 0·4	+ 0·1

	May	June	July	Aug
Westerly	− 3·7	+ 0·4	− 0·7	− 10·4
Cyclonic	+ 3·6	+ 1·0	− 1·3	+ 1·5
Anticyclonic	− 2·6	− 2·0	+ 4·0	+ 0·3
North-westerly	+ 1·0	− 3·2	+ 2·4	+ 1·1
Easterly	+ 0·2	+ 4·1	− 0·5	+ 0·8
North-easterly	− 0·2	− 1·4	− 1·8	+ 2·2
South-easterly	+ 0·2	+ 0·9	− 0·5	+ 0·9
Northerly	+ 2·2	− 2·0	+ 1·4	+ 3·6
South-westerly	+ 0·2	− 0·5	− 2·0	− 1·8
Southerly	− 0·2	+ 0·5	− 0·7	+ 1·5
Unclassifiable	− 0·7	+ 2·2	+ 0·3	+ 0·5

	Sept	Oct	Nov	Dec
Westerly	+ 1·0	+ 1·4	− 6·4	− 7·0
Cyclonic	+ 2·9	− 2·7	+ 1·3	− 3·5
Anticyclonic	− 6·8	+ 1·8	+ 3·3	+ 0·2
North-westerly	− 0·4	+ 2·5	+ 1·6	+ 3·6
Easterly	− 0·3	− 2·5	+ 2·5	+ 0·7
North-easterly	+ 1·1	− 1·9	+ 0·3	+ 1·8
South-easterly	+ 0·6	− 0·5	− 0·7	+ 1·0
Northerly	− 3·7	+ 0·1	− 2·7	+ 4·0
South-westerly	+ 0·2	+ 0·4	+ 1·7	− 3·4
Southerly	+ 2·8	+ 0·7	− 0·6	+ 1·0
Unclassifiable	+ 2·6	+ 0·7	− 0·3	+ 1·6

caused the cancellation of the Scottish ski championships. Records for Glencoe suggest that in some years during the last decade nearly 50 per cent of all days out of a season of approximately 120 days were described by the chairlift operators as being unsuitable or very unpleasant for sport.

Recent Climatic Changes

These high figures have occurred at a time when general changes in the frequency of occurrence of the main weather types (Lamb, 1972) have resulted, as shown in Table 10.1, in less westerly type, particularly in winter, and more northerly, easterly and anticyclonic weather.

Unpublished work by R. Blackmore (1971) suggests that these changes should lead to increased accumulation of snowfall and a reduction in ablation amounts, but the monthly data probably conceal changes of singularity occurrence and pattern which Finch (1971) suggests may have been important in mid-winter months in recent years. A preliminary analysis of the occurrence of the major type-groups during the first ten pentads of the year (Table 10.2) does suggest that a steep rise in the frequency of southerly types at the end of January has occurred.

Table 10.2

FREQUENCY OF OCCURRENCE IN DAYS OF MAJOR TYPE-GROUPS DURING
FIRST TEN PENTADS OF THE YEAR (BASED ON DATA FOR 1949–70)

Type	Pentad Number									
	1	2	3	4	5	6	7	8	9	10
Westerly	28	31	36	28	19	34	35	31	19	25
Anticyclonic	21	12	17	16	27	15	20	24	7	18
Northerly	21	16	12	8	7	3	7	13	18	14
Easterly	10	9	11	7	4	7	6	6	17	15
Southerly	3	8	3	13	7	21	7	6	8	6
Cyclonic	13	13	13	12	14	8	10	15	14	9

Studies of ablation rates suggest that the advection of mild air in winter is more effective than high radiation values in removing snow-cover and therefore the forecaster must pay special attention when there seems a likelihood that maritime tropical air from the south-west will envelop Scotland. If, at the same time, pressure is high over the Continent, leading to further warming by subsidence, local *föhn* effects can lead to high temperatures in the Cairngorms, as happened in January 1971, when Glenmore Lodge (323·1 m) recorded a maximum temperature of 16°C.

The foregoing paragraphs have highlighted only a few of the relationships between climatic and economic factors in this particular enterprise. The number and type of climatic problems that the industry faces should discourage investment in costly capital equipment, and it may well be that the viable exploitation of the area is to be found in projects which involve smaller capital outlay and are flexible to operate, being easily shut down or opened up as opportunity demands.

Acknowledgements

The author would like to thank the Meteorological Office (for supplying the weather-type data), and the Glasgow Weather Centre, Scottish Council for Physical Recreation, Cairngorm Sports Development Ltd, and White Corries Ltd (for providing information about the skiing industry).

References

BLACKMORE, R. (1971). *Personal communication.*
FINCH, C. R. (1971). The changing face of January in S.E. England. *Weather*, **26**, 2–6.
GREEN, F. H. W. (1968). Persistent snowbeds in the western Cairngorms. *Weather*, **23**, 206–9.
LAMB, H. H. (1969). The new look of climatology. *Nature, Lond.*, **223**, 1209–15.
LAMB, H. H. (1972). British Isles weather types—a hundred years' register of the sequence of circulation patterns. *Met. Office, Geophys. Mem.* (1972).
SCOTTISH DEVELOPMENT DEPARTMENT (1967). *Report on Cairngorm area.* HMSO, Edinburgh.
SPINK, P. C. (1968). Scottish snowbeds in summer 1967. *Weather*, **23**, 209–11.
SPINK, P. C. (1970). Scottish snowbeds in summer 1969. *Weather*, **25**, 201–4.
UNIVERSITY OF EDINBURGH (1970). *Impact of tourist development in Upper Speyside.* Department of Geography, University of Edinburgh.

CHAPTER ELEVEN   A. H. PAUL

Weather and the Daily Use of Outdoor Recreation Areas in Canada

Weather conditions affect the pursuit of many human activities. Leisure-time activities are no exception. This chapter reports a study of weather influences on day-to-day participation in summer outdoor recreation in Canada. Despite the expanding body of research in the general area of weather effects on human activities, recreation has received very little attention (Maunder, 1970). Only recently has any detailed research been conducted in this field. Dowell (1970) has completed a study relating weather to daily launchings of pleasure boats on an Arkansas reservoir. Maunder (1970) reported an examination of the occupancy records of Lake Louise camp site, Banff National Park, relative to weather; results were inconclusive. Two further projects are under way in the United States at present, one at the Department of Park and Recreation Resources, Michigan State University, and the other at the Center for Recreation Resources Development, University of Wisconsin, Madison.

Approach
Eight outdoor recreation activities were analysed within three widely separated and climatically diverse regions of Canada (Fig 11.1), viz, Edmonton, Alberta; Ottawa-Hull; and southern New Brunswick. The eight activities are swimming, beach-use, visiting multi-activity parks, picnicking, boating, driving for pleasure, outdoor sports, and visits to outdoor sites of special interest such as zoos and historical exhibits. Daily attendance records for the 1969 summer, defined as 15 May to 15 September inclusive, were collected for many outdoor recreation facilities in each study region. These records were subjected to computer analysis together with the climatological observa-

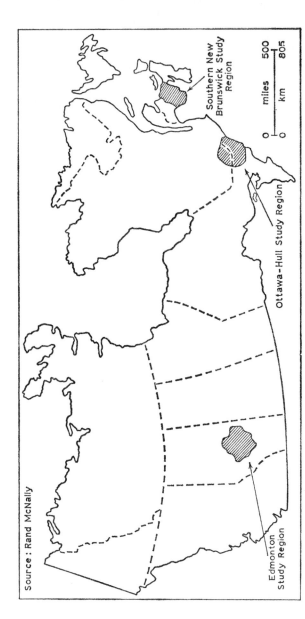

Fig. 11.1 Location of the three study regions

tions routinely stored on punched cards by the Canadian Meteorological Branch.

Four sets of weather circumstances to which the recreationist may react can be distinguished. These are (a) on-site weather, (b) conditions at the trip origin, (c) the forecast, and (d) conditions as anticipated by the recreationist, whose perception of weather is less than perfect. The present study concentrates on the first two sets of circumstances, or the actual weather.* Weather variables selected for study were daily maximum temperatures, total daily and six-hourly afternoon precipitation, mean daily relative humidity, mean windspeed over the period 0900–1900 hours local time, total daily sunshine hours, and sunshine hours experienced between 1100 and 1700 hours local time. Values of two 'comfort indices', Thom's Discomfort Index and Effective Temperature, calculated for 1500 hours local time, were also employed. These indices combine two or more meteorological elements in a single numerical variable. Temperature and relative humidity are incorporated in Thom's Discomfort Index, while Effective Temperature includes these two elements together with windspeed.

Limiting the period studied to one summer (1969) has the advantage that many of the variables in the recreational system, such as changes in transportation systems, provision of new parks and recreation facilities, growth in population, and many others, remain essentially constant during the period, and can therefore be ignored. Two factors influencing recreation participation which must be taken into account in this analysis, however, are day of the week (weekday or weekend) and stage of the summer (school vacation or not). To facilitate the identification of day-to-day fluctuations in attendance due to weather, the 124-day summer was subdivided into six day-groupings as shown in Table 11.1, and data within each day-grouping were investigated separately.

The sixfold subdivision of the summer season results in some small samples of data for study, especially in day-groupings (iv) and (v). And, in

---

* Future research might fruitfully consider the other two sets. Indeed, the effect of the forecast is currently under review at Michigan State University. (*Personal communication*, letter from Mr D. M. Crapo, Department of Park and Recreation Resources, Michigan State University, East Lansing, Michigan, 28 April 1970.)

general, the data collected do not provide an adequate basis for rigorous statistical analysis. Thus the primary output from the computer is simply a plot or scattergram of facility attendance (Y) against a single weather variable (X). (A multiple regression approach, with facility attendance as the dependent variable, might be taken although the weather variables selected for analysis are not independent of one another.) Since the study is of a basic

Table 11.1

SUMMER DAY-GROUPINGS

Day-Grouping Number	Number of Days	Day	Period
Stage of Summer			
A School-Vacation Period:			
(i)	13	Sundays & Statutory Holidays	
			28 June–
(ii)	10	Saturdays	1 September
(iii)	43	Weekdays	
B Summer outside School-Vacation Period:			
(iv)	9	Sundays & Statutory Holidays	15 May–27 June
			and
(v)	8	Saturdays	2–15 September
(vi)	41	Weekdays	

The Statutory Holidays, equivalent to English Bank Holidays, are Monday, 19 May; Tuesday, 1 July; Monday, 4 August; and Monday, 1 September, 1969.

and exploratory nature, the interpretation of the data plots provides valuable initial insight into recreation–weather relationships. For each plot a linear correlation and regression analysis was performed. A run-test on the deviations from the regression line provides a statistical measure of the validity of linearity in the relationship (Crow, Davis, and Maxfield, 1960). If linearity is rejected, the scattergram indicates the kind of curve-form which should be proposed.

Activity	Maximum Temperature	Precipitation	Sunshine Hours	Wind	Relative Humidity	Comfort Indices
Swimming	staight line often with high correlation	decay curve	straight line, lower correlation than max. temperature	weak inverse relation	straight line, variation of r-values great	straight line often with high correlation
BeachUse	straight line with fairly high correlation	decay curve	straight line, higher correlation than max. temperature	weak inverse relation	straight line, great variation of r-values	straight line with fairly high correlation
Multi-Activity Parks	low linear r-values; some rejection of linearity	very shallow decay curve	straight line, weak positive relation	almost no relation	inverse relation, linearity not rejected	very similar to relation for max. temperature
Picnicking	very little association; no rejection of linearity	straight line, weak inverse relation	almost no association	almost no relation	weak inverse linear relation	very little association

(continued below)

Activity	Maximum Temperature	Precipitation	Sunshine Hours	Wind	Relative Humidity	Comfort Indices
Boating	very little association	difficult to identify	almost no relation	weak inverse relation	almost no relation	little association
Pleasure Driving	slight positive relation	almost no relation	almost no relation	weak inverse relation	difficult to identify	slight positive relation
Visiting Special Sites	bell-shaped curve	shallow linear decay	straight line, moderate correlation	weak inverse relation	inverse linear relation	bell-shaped curve
Outdoor Sports	bell-shaped curve	decay curve	straight line, varying r-values	inverse relation	straight line, fairly high negative correlation	bell-shaped curve

Source: Computer analyses, 1970

Fig. 11.2 **Summary of general relationships found between activity participation and weather variables**

Results

A general summary of the relationship between daily participation in the eight outdoor activities and the weather elements analysed is given in Fig 11.2. The influence of weather on daily participation varies from activity to activity. Swimming and beach-use demand very specific weather conditions. Other pursuits such as picnicking and pleasure driving are engaged in over a broader range of weather situations.

Attendances at swimming areas are strongly influenced by daily weather conditions. Maximum temperature is the atmospheric variable most closely associated with numbers swimming: higher temperatures attract more swimmers. The general relationship between swimmer attendance and maximum temperature can be considered as a straight line; high correlation coefficients are generally found. Daily precipitation, mean relative humidity, and mean windspeed are inversely related to swimming activity, but the relationships appear to have different forms, as shown in Fig 11.2. Sunshine hour totals have a positive linear association with daily participation in swimming, but correlation coefficients are lower than for daily maximum temperature. The relationships of the two comfort indices to swimming participation are similar to that of daily maximum temperature, but the correlation coefficients using these indices are slightly lower than for temperature alone.

Regional variation in the response of swimmers to the weather is quite apparent. This is manifested mainly in analyses involving temperature, sunshine, and relative humidity, with windspeed and precipitation responses showing little geographical variation. Temperature will be considered as an illustration. In midsummer (the school vacation), daily maxima must exceed 22°C before swimming facilities in the Ottawa-Hull region receive any appreciable patronage (defined as 10 per cent of the peak attendance recorded). The comparable value in both the Edmonton and southern New Brunswick study regions is 19°C. It is noteworthy also that swimmers in all three regions will tolerate temperatures 3–4°C lower in May, June, and September than during the school vacation period, when temperatures are usually higher.

Beach-use is defined as 'beach activity of people or individuals including non-swimmers and children'. Thus wading, swimming, sunbathing, ball games, surfboarding, bird watching, nature study, and all other activities which take place on the beach are included. Beach-use is subject to weather influences not markedly different from those characteristic of swimming. Maximum temperature and sunshine hour total are the variables most closely related to beach-use, as is the case with swimming. For both activities, precipitation amount is related to attendance by a decay curve (Fig 11.2), but that for beach-use is shallower. Maximum temperatures 3–6°C lower than for swimming will be tolerated by beach users, but the relationship between use and temperature is again linear. Correlation coefficients, however, tend to be somewhat lower for beach-use than for swimming.

'Multi-activity parks' are those less-developed parks which allow participation in a broad range of outdoor pursuits. They are difficult to define in precise terms. They are not wilderness areas, but neither are their amenities developed to the extent that they are well known for one particular activity, as is the case with the beach park. Visitors are attracted by the potential for activities such as fishing, boating, camping, and nature study, and by the general outdoor environment, rather than by the mass-participation forms of outdoor recreation such as swimming. Weather has only a modest influence on daily attendance at multi-activity parks (see Fig 11.2 for details). Low correlations between weather and visitation are easily explained at this type of facility. Within the parks themselves, visitors will tend to select the activity most suited to the current weather conditions.

Picnicking and pleasure driving are two activities on which weather appears to have little influence. Daily use of picnic areas is directly but weakly related to maximum temperature; there is a slight inverse relation with both precipitation amount and mean relative humidity. Participation in driving for pleasure has a low positive association with daily maximum temperature, and even less correlation with the other weather variables. Perhaps more interesting is the fact that it is the only one of the eight activities studied which is apparently unaffected by the occurrence of precipitation.

Surprisingly, the intensity of boating activity was found to bear little relation to weather conditions. However, the analysis was restricted to pleasure boating records from lock-stations on the Rideau Canal, which links Ottawa and Kingston, Ontario. A wider sample might perhaps yield more significant results. Furthermore, precipitation observations in close proximity to the Canal were so few in 1969 that it proved impossible to draw any conclusions concerning this variable. The slight positive correlation found between day-use boating and maximum temperature, together with the inverse relation with mean windspeed, agrees with the findings of Dowell (1970). He

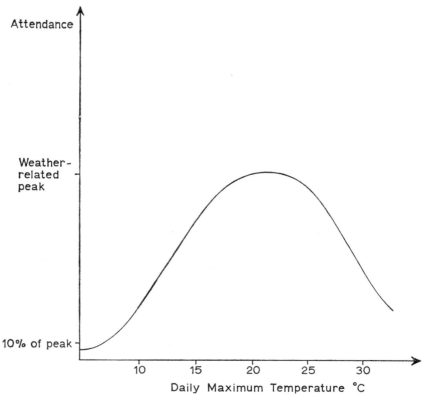

Fig. 11.3 Schematised relation between visitation to a special site and daily maximum temperature

investigated boat-launchings on an Arkansas reservoir with respect to four meteorological parameters and found somewhat stronger relationships to exist than those identified in the present study.

Two pursuits which have largely been ignored in the past in studies of outdoor recreation complete the eight activities analysed. These are: visiting special sites and participation in outdoor sports. The special sites examined in the project include two zoos, four historical exhibits, one wildlife sanctuary, and one site of outstanding geological interest. Outdoor sports analysed include golf and tennis, besides participation in the sports programmes of a number of New Brunswick parks. Some very distinctive relationships between participation in these two broad groups of outdoor activity and weather were identified.

The curve in Fig 11.3 represents the association between daily attendance and maximum temperature found to be characteristic of both visitation to special sites and outdoor sports. While there is a strong relationship, it is anything but linear, showing a definite bell shape. Precipitation amount is inversely associated with numbers engaged in either activity, but its importance varies greatly. It is only slightly, and apparently linearly, associated with visitation to special sites, but has a very strong deterrent effect on participation in sports programmes and tennis. The single weather variable most closely related to attendance at special sites is total daily sunshine hours. On the other hand, the greatest influence on rounds of golf played was found to be that of mean relative humidity. Tennis-playing is affected most by precipitation.

It is evident from the few diverse special sites and outdoor sports reviewed that weather effects on attendance are markedly different from those pervading the activities discussed earlier. New variables apparently move to the forefront of the total weather situation. The dominant position of maximum temperature has receded. While maximum temperature alone may be used as a reasonable predictive variable for participation in activities such as swimming and beach-use, for others, and particularly for outdoor sports, a wider range of weather elements must be considered.

Discussion

After this brief review of the effects of weather on individual activities, it is time to examine the role of weather in outdoor recreation behaviour in general. To facilitate this study, a series of maps was drawn. These maps show the relative popularity of all facilities within each study region on a number of days during the 1969 summer. The days selected were illustrative of varying weather conditions. Attendance at each facility was expressed as a percentage of the peak daily value recorded during the entire summer. The maps are thus based on the spatial distribution of recreationists, and they show some very interesting features. Only one illustration is presented here, but many similar occurrences were found.

On 5 July 1969, Edmonton had a maximum temperature of 17°C and 1 mm of precipitation. There were 3·9 hours with bright sunshine. Attendances at all swimming facilities and beach parks in the Edmonton study region were very low. None of the outdoor swimming places received more than 2 per cent of its peak use. The six beach parks fared slightly better than this, one recording a use of 19 per cent of peak level, but four of the other five experienced 7 per cent or less. On the other hand, Edmonton's Riverside Golf Course was relatively popular. Rounds played were 62 per cent of the total on the busiest day of the summer.

On 10 August 1969, the maximum temperature at Edmonton was 26°C and there were 13·1 hours of bright sunshine. A brief evening thundershower resulted in 9 mm of precipitation. Almost all outdoor recreation facilities studied in the region had more visitors than on 5 July. The significant difference, however, is the relative popularity of the different types of recreation area. The only swimming site outside Edmonton City and four of the six beach parks recorded their peak attendances for 1969. Swimmers at pools in the city of Edmonton, however, ranged from only 32 to 66 per cent of the 1969 daily maximum. The number of rounds played at Riverside Golf Course was actually lower on 10 August than on 5 July.

It is thus clearly demonstrated in the series of maps that the weather situation does more than merely persuade people to take part or not in outdoor recreation. Certainly warmer weather is instrumental in more people choosing to spend time out-of-doors. But weather also plays a major role in shaping the choice of activity and the location of the recreational facility to be used. The maps strongly support the conclusion, from the activity analyses reported earlier, that some outdoor pursuits are more influenced by weather conditions than others. A hierarchy of relationships between activity participation and the more significant weather variables may be developed. In most instances, maximum temperature is the single variable most closely related to facility use. An hypothesis for the choice of a recreational activity under different temperature conditions is proposed in Fig 11.4. This diagram is a schematic illustration of the 'weather-selective' behaviour of outdoor recreationists. Daily maximum temperature, although important, is of course only one weather variable. The total weather situation must be considered in developing a complete hypothesis of activity selection.

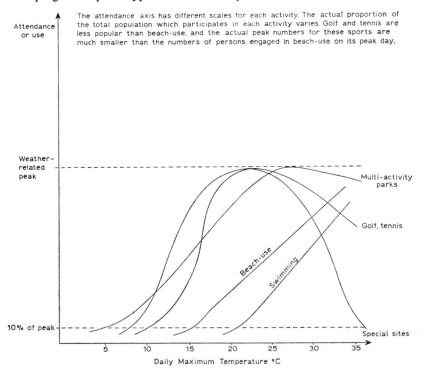

Fig. 11.4 Illustration of the weather-selectiveness of outdoor recreationists

Figure 11.4 depicts a simplified situation where a choice is made between only five forms of outdoor recreation. The relationships between participation in the individual activities and maximum temperature have been plotted on one set of axes. With maxima below 16°C, only multi-activity parks, golf and tennis, and special sites receive any significant patronage. These activities are still most favoured between 16 and 21°C, but in this range of maximum temperature, beaches and swimming areas also receive some use. Use of golf courses, tennis courts, and special sites peaks at about 24°C, and that of multi-activity parks at about 27°C. Beach-use and swimming are selected by increasing numbers of outdoor recreationists above 21°C, and above 27°C are strongly favoured activities. A decrease in use of outdoor sports facilities, special sites, and multi-activity parks sets in as temperatures climb well above 27°C and other outdoor activities become increasingly popular.

Weather-selectiveness has a striking effect not only on activity choice, but also on the outdoor recreationist's choice of location on given days of the summer. The fact that certain pursuits are favoured on certain days will result in locations with the best potential for these pursuits being preferred also. This has obvious implications concerning the flow of recreational traffic on highways. Weather-selectiveness also has an important implication for the commercial operator of outdoor recreation facilities. The facility with a range of outdoor activities favoured under the widest range of weather conditions has a distinct competitive advantage over one with only a single attraction, such as the historical exhibit or the beach.

The outdoor recreationist, and the facility manager especially, would be greatly aided by a weather forecast which emphasised the weather parameters most pertinent to the recreation industry. These parameters, besides the current emphasis on temperatures, might include the approximate timing and likely duration and amount of precipitation, the total daily sunshine hours, and the mean daily relative humidity.

The incorporation of weather parameters in a model for predicting daily attendance at outdoor recreation facilities would be most valuable. If a prediction technique is to be of value to the facility manager, it must be quick to apply and simple to use. Swimming may be employed as an example, using maximum temperature as the weather variable. The straight-line relationship fits fairly well over the range of temperatures normally experienced. With high linear correlation coefficients, accurate predictions can be rapidly made from maximum temperature, day of the week, and stage of the summer. Ratios of standard error of estimate to mean attendance are as low as 1 : 3. Assuming normal distribution of deviations from the regression line, then 68 per cent of all predictions based on maximum temperature alone will be within 33 per cent of the actual attendance. The addition of other information such as expected precipitation or windspeed will improve the prediction still further. The three variables: day of the week, stage of summer, and one relevant weather parameter, are sufficient to allow useful prediction of daily attendances only at swimming facilities, beaches, and outdoor sports areas. More complex procedures would be required at other types of facility, involving detailed considerations of two or three further weather variables.

The knowledge generated in this study on the nature of weather-recreation relationships may be used in defining the climatic capability of a given area to support particular outdoor recreation activities. It is possible to state for any locality the number of days during the year on which weather conditions should fall within the tolerable range for a given activity. Such information can form an input to the second stage of the Canada Land Inventory's classification system of land for recreation (Canada, Department of Forestry and Rural Development, 1967). At present this classification emphasises land and water resources rather than the total environment. The desirability of including weather and climate within the framework of such a system has been explicitly stated by the Outdoor Recreation Resources Review Commission (ORRRC, 1962).

Conclusion and Recommendations
This study has explored the nature of the relationships between weather and participation in a number of outdoor recreation activities. The role of weather in the much-analysed process of selection of activity and destination by outdoor recreationists has also been reviewed. Further details can be obtained from the full report of the investigation (Paul, 1971). Several potential practical applications have been identified, but further study and a larger body of high-quality recreational and meteorological data are required before these applications can be fully realised.

Analysis was confined to the relationship between recreation participation and weather conditions. Future studies should also examine the influence of the weather forecast on outdoor recreation behaviour. The recreationist's interpretation of the forecast, and, indeed, even his frequent ignorance of it, may have a marked bearing on his decision-making. Efforts should also be directed towards a multivariate analysis of the weather-recreation interaction. Although several weather variables were investigated in this study, they were tested separately. More sophisticated analyses should encompass a multi-factor approach, but such a treatment must await the standardisation and improvement of recreational data collection. These data must also be stored and tabulated in a systematic and useful format. Nationwide standards should be set for these procedures as is the case with weather observations. When more specific recreational data are available for a period of several years, more complete statistical analysis will be possible. Finally, an extension of the study to other outdoor pursuits would provide information useful to the recreation administrator regardless of the activity of his concern. With an initial approach already formulated, such an extension is a logical complement to the project reported here.

References

CANADA, DEPARTMENT OF FORESTRY AND RURAL DEVELOPMENT (1967). *Field Manual —Land Capability Classification for Outdoor Recreation*. ARDA Canada Land Inventory, Ottawa.

CROW, E. L., DAVIS, F. A., and MAXFIELD, M. W. (1960). *Statistics Manual*. Dover Books, New York.

DOWELL, C. D. (1970). *The Relationship of Reservoir Pleasure Boating to Selected Meteorological Factors*. Unpublished PhD thesis, Texas A & M University.

MAUNDER, W. J. (1970). *The Value of the Weather*. Methuen, London.

ORRRC (1962). *Alaska Outdoor Recreation Potential*. Study Report 9, Outdoor Recreation Resources Review Commission, Washington, DC.

PAUL, A. H. (1971). *Relationships of Weather to Summer Attendance at Some Outdoor Recreation Facilities in Canada*. Unpublished PhD thesis, University of Alberta, Edmonton, Alberta.

CHAPTER TWELVE   J. D. McQUIGG

Simulation Model Studies of the Impact of Weather Factors on Road Construction and the Movement of Heavy Equipment in Agricultural Operations

Introduction

The management-decision problem described in this chapter is real. Evaluation of the alternative courses of action open to farm managers faced with this problem could certainly be made in a more straightforward manner, given a large body of real operational records and experience over a sufficiently long period of time. In this particular instance, a long record of real field work experience is not available. However, excellent long-term daily weather observations are readily available for application in a simulation model, and the results lead to a reasonable solution of the problem.

The General Problem

A serious infestation of Southern Corn Leaf Blight occurred (US Dept of Commerce *et al*, 1970) in most of the corn (maize) producing areas of the eastern United States in 1970. Several factors contributed to this situation: (a) a new race (Race T) of the pathogen (*Helminthosporium maydis*), which is especially virulent on corn hybrids containing a male sterile cytoplasm, (b) widespread use of hybrids susceptible to the new race, and (c) favourable weather conditions.

One course of action which at first appeared to be open to farm managers as the 1970 autumn corn harvest was about to begin was to attempt to complete harvest operations as soon as possible, and then to plough in the crop residue, in the hope that this would reduce the likelihood of the pathogen remaining alive through the winter. This was possible in some areas of Missouri, but by 12 December 1970 only two-thirds of autumn ploughing was completed (USDA *et al*, 1970) because of abundant autumn rains. Another course of action available to corn farmers as the 1971 crop season began was to plant blight-resistant hybrid seed corn varieties. However, only a portion (22 per cent) of the seed available for the 1971 crop was from normal cytoplasm stock (USDA, 1970). Many farmers had to plant seed varieties that were not known to be blight-resistant. Some farm managers could choose to plant alternative crops, such as soybeans or grain sorghums. However, with good prospects for favourable corn prices in 1971, many farmers were willing to assume a fairly high degree of risk of blight damage and chose corn as their major crop.

At mid-winter 1971, there was one important course of action still definitely open. This was to attempt to plant the 1971 corn crop early. According to Dr Wm. J. Murphy, University of Missouri Agronomy Specialist, 'Early planting does not affect the susceptibility of corn varieties to the disease. But early-planted corn is likely to be farther along in development when the blight does strike, thus reducing detrimental effects from loss of photosynthetic surface in the leaves, and offering less chance for ear and stalk rot to occur.' Dr Murphy also stated, 'It would appear that if a Missouri corn producer is unable to get corn planted by late April in extreme southern Missouri, or by May 15–20 in central and northern Missouri, he should then strongly consider going to an alternative crop in 1971.'

The Specific Problem

Since substantial amounts of autumn ploughing remained uncompleted at the end of 1970, many farm managers, who planned to plant their 1971 corn acreage early, needed to assess the likelihood of completing ploughing, fertiliser application (if any), seed bed preparation, and planting, in a limited number of favourable working days. The discussion that follows concerns the preparation and use of simulated field work 'experience', based on a long series (1918–65) of daily temperature, snowfall, and precipitation observations.

The Simulation Model

Maunder, Johnson, and McQuigg (1971) used a four-year series of daily operational records from a major road construction project in central

Missouri to develop an algorithm that transforms daily rainfall observations into a series of numbers that closely resemble road-building operational records. The chief advantage of this procedure is that there are many locations for which at least thirty years of daily weather records are readily available. Application of the algorithm, then, produces what are believed to be realistic estimates of the average amount of time that road-building (agricultural) equipment can be operated at certain periods of the year. More important, the simulated operational data, based on longer periods of record than available actual operational data, provide useful estimates of variation about the average values.

Lack of space prevents detailed description of the algorithm. In brief, it is based on estimates of loss of water from the surface layer of a bare soil (Forest Service, USDA *et al*, 1959), and constants which reflect the soil type (silt, sand, or clay). The actual operational records maintained by the highway commission resident engineer on the road-building project classified each day as:

> a holiday or weekend, type '0',
>
> a full work-day, type '1',
>
> a no-work day, type '2', or
>
> a partial work-day, type '3'.

The algorithm separates the simulated days into these same categories. A detailed description of this procedure is given in Maunder, Johnson and McQuigg (1971).

Application of the Algorithm to the Spring Planting Problem

At the 1970 autumn conference of University of Missouri agronomists it was proposed to apply the road-building simulation model to the problem facing Missouri farmers during the 1971 planting season. This suggestion met with an enthusiastic response. In early December preliminary results from use of the model were discussed in a meeting with about 200 corn farmers from a four-county area in north-central Missouri. The material that follows is based on that presentation. (Presentations were also made by University staff from Agronomy, Plant Pathology, Agricultural Engineering, and Agricultural Economics.)

Simulated Field Work Data for Brunswick, Missouri

The operational road construction data for central Missouri were from an area in which the major soil type is silt loam. Assuming that the soil water-holding characteristics of the top layer of silt loam soil near Brunswick were not substantially different then direct application of the model for that soil type was permissible, with the major difference in results being related to different rainfall patterns at Brunswick.

Consultation with members of the Agronomy Department of the University of Missouri led to modification of some of the constants in the algorithm, and preparation of simulated field work data for clay and sandy soils, both of which are also prevalent in the Brunswick area.

In most years field work begins to be active about mid-April in the northern sections of Missouri. Allowing a value of 1·0 for a simulated 'full-work' day, 0·5 for a 'partial-work' day, and 0·0 for 'no-work' day (including holidays and weekends as potential work days), we obtained the results summarised in Table 12.1.

Table 12.1

BRUNSWICK, MISSOURI, NUMBER OF DAYS' WORK COMPLETED BY MAY 15

Desired Starting Date	Poorest Year in Twenty			Average Year			Best Year in Twenty		
	Sand	Silt	Clay	Sand	Silt	Clay	Sand	Silt	Clay
March 15	32	21	16	42	34	26	52	47	36
April 15	16	10	8	21	16	13	26	22	19

Many northern Missouri farmers were hoping to start field work earlier than April 15 in 1971. The model was next applied to Brunswick meteorological records, assuming a desired starting date of March 15, and adding constraints related to temperature and snow cover. In addition to a consideration of soil moisture, if the daily mean temperature was −2·25°C, or lower, or if there was 2·5 cm or more of snow on the ground, the day was classed as a 'no-work' day. Results of this application are also shown in Table 12.1.

Conclusion

The values shown in the body of Table 12.1 can be viewed as specialised climatology. The units used are readily interpreted by a particular class of decision-maker. The decision problem facing farmers in the corn belt of the United States prior to planting is highly complex. The amount of time available for land preparation, planting and early cultivation is only one of a number of related factors that must be evaluated.

Table 12.2

SILT LOAM SOIL NEAR BRUNSWICK, MISSOURI: LINEAR PROGRAMMING
SOLUTIONS

Number of Days' Work Done Prior to 15 April	Probability Level for Work 15 April–15 June	Estimated Net Income $	Value of an Extra Work-Day ($)			
			15–30 April	1–15 May	15–31 May	1–15 June
30	5%	30 620	343	0	0	0
30	50%	31 306	343	0	0	0
30	95%	31 544	0	0	0	0
0	5%	18 026	978	978	978	610
0	50%	25 237	978	978	978	610
0	95%	28 901	0	0	0	0

Following the four-county meeting in early December 1970, Professor Herman Workman (University of Missouri Agricultural Economics Department) proposed that the simulated work-day values be used in a linear programming solution. The result is a procedure in which a farm manager coming to a University of Missouri Extension Center is asked to provide information concerning machinery, land, labour, the periods of time in which he desires to complete land preparation, planting and early cultivation, and such factors as cost of seed, fertiliser, expected yields and prices, etc. Three solutions are provided (for about $10.00 worth of computer time) for a major soil type, ie, for the poorest work-year in twenty, for an average work-year and for the best work-year in twenty. (Different probability levels can be readily substituted.)

M.U. CROP PLAN NO. 1   DATE: 01/05/71
FOR TOM JONES

TABLE 1: INFORMATION USED IN THIS CROP PLAN.
CROP MACHINERY INCLUDES 2 TRACTORS

FIELD TIME PERIODS SPECIFIED:

P1 PREPARING LAND BEFORE PERIOD 1
P2 APRIL 15 TO APRIL 31
P3 MAY 1   TO MAY 15
P4 MAY 16   TO MAY 30
P5 JUNE 1   TO JUNE 15
P6 PLANTING CROPS AFTER PERIOD 5

FIELD OPERATION RATES SPECIFIED:

PREPARATION OF LAND....... 20.6 ACRES PER DAY
PLANTING CROPS................ 65.0 ACRES PER DAY
CULTIVATING CROPS........... 70.0 ACRES PER DAY

ESTIMATED FIELD TIME AVAILABLE:

P1 GOOD WORK DAYS........ 30.0
P2 GOOD WORK DAYS........ 5.0
P3 GOOD WORK DAYS........ 5.0
P4 GOOD WORK DAYS........ 5.0
P5 GOOD WORK DAYS........ 5.0

LAND AND CROP LIMITS SPECIFIED:
MAXIMUM CROPLAND.......... 500.0 ACRES

TABLE 2: SUMMARY OF THIS CROP PLAN.

CROP & PERIOD PLANTED	ACRES	CROP COST PER ACRE	YIELD PER ACRE
CORN P2...............	325.0	$50.00	90.0 BU.
SOYBEANS P5........	175.0	$30.00	35.0 BU.

PRODUCTION	AMOUNT	VALUE PER UNIT
CORN BUSHELS.....	29250.0	$1.30
SOYBEANS BUSHELS	6125.0	$2.60

Fig. 12.1 A sample print-out of the linear programming solution provided for farm managers (*continued on page opposite*)

TABLE 2: SUMMARY OF THIS CROP PLAN. (CONT.)

COSTS AND RETURNS	AMOUNT

GROSS VALUE OF CROPS PRODUCED............ $53950.00
LESS SPECIFIED CROP & MACHINERY COSTS. $23330.35

NET VALUE OVER SPECIFIED COSTS.............. $30619.65

TABLE 3: CROPS NOT PLANTED IN THIS PLAN.

CROP & PERIOD PLANTED	CROP COST PER ACRE	REDUCED COST NEEDED
CORN P3.............................	$50.00	$ 6.23
CORN P4.............................	$50.00	$20.00
CORN P5.............................	$50.00	$33.00
SOYBEANS P4......................	$30.00	$.
SOYBEANS P6....................	$30.00	$22.80
MILO P4.............................	$40.00	$13.00
MILO P5.............................	$40.00	$24.00
MILO P6.............................	$40.00	$31.80

TABLE 4: VALUE OF FIELD TIME PER DAY.

PERIOD	MAXIMUM AVAILABLE	AMOUNT USED	VALUE OF AN EXTRA DAY
P1 FIELD TIME.......	30.0 DAYS	18.9	$.
P2 FIELD TIME.......	5.0 DAYS	5.0	$343.57
P3 FIELD TIME.......	5.0 DAYS	5.0	$.
P4 FIELD TIME.......	5.0 DAYS	5.0	$.
P5 FIELD TIME.......	5.0 DAYS	2.7	$.

TABLE 5: VALUE OF FIELD OPERATIONS PER ACRE.

OPERATION	VALUE OF AN EXTRA ACRE COMPLETED
P1 PREPARE LAND.	$ 2.43
P2 PREPARE LAND.	$19.11
P3 PREPARE LAND.	$ 2.34
P4 PREPARE LAND.	$ 2.43
P2 PLANT CROP.....	$ 6.05
P4 PLANT CROP.....	$ .77
P5 PLANT CROP.....	$ .77

A sample print-out of the linear programming solution which is provided to the farm manager is shown as Fig 12.1. A summary of results including this and five other possible situations is shown as Table 12.2. (The values for land, machinery, costs, etc, are the same as shown in Fig 12.1.)

The computer cost for one solution was less than $2.00. Thus, for a very small sum, and a bit of effort in assembling machinery, land, yield estimates, and costs for a particular farm, it is possible to evaluate several alternatives, such as a shift to an alternative crop, buying or leasing added machinery, allowing for a reasonable range of variation in the amount of work-time available during critical periods of time.

The effort to publicise the alternatives available to corn farmers during the winter months prior to the 1971 crop season, plus a highly favourable spring season, resulted in early completion of the planting operation in Missouri (and in many other sections of the Corn Belt). This in turn resulted in earlier maturity of the crop. Comparatively dry, cool, late summer weather reduced the damage from blight, with the result that the 1971 corn crop was very large.

References

FOREST SERVICE, USDA and US ARMY ENGINEERS WATERWAYS EXPERIMENT STATION (1959). *Development and Treating of Some Average Relations for Predicting Soil Moisture*. Tech. Memo 3–331, Report No 5, Vicksburg, Mississippi.

MAUNDER, W. J., JOHNSON, S. R., and McQUIGG, J. D. (1971). A Study of the Effect of Weather On Road Construction: A Simulation Model. *Mthly Weather Rev.*, **99** (12).

SMITH, C. V. (1970). Weather and Machinery Work-days, in *Weather Economics*. Ed. J. A. Taylor, Pergamon Press, Oxford, 17–26.

US DEPARTMENT OF AGRICULTURE (1970). *Statistical Reporting Service*, SeHy 1–4 (11–70) Exp. Supply of S.C. Hyb. Grain Sorg. Seed, Washington, DC, 19 November.

US DEPARTMENT OF AGRICULTURE and US DEPARTMENT OF COMMERCE (1970). *Crop Weather*, 14 Dec. Columbia, Missouri.

US DEPARTMENT OF COMMERCE and USDA (1970). *Weekly Weather and Crop Bulletin*, **57**, (52) 11–14, 28 Dec.

CHAPTER THIRTEEN   J. S. HAY
                                              and C. P. YOUNG

Weather Forecasting for the Prevention of Icy Roads in the United Kingdom

Introduction

In Great Britain and many other countries the most common way of preventing the accumulation of snow and ice on roads is to spread salt. The amount of salt being used for this purpose has risen by a factor of about six over the past fifteen years and consumption in the United Kingdom is now running at the rate of about 1·5 million tons per year. Demand has so outstripped the domestic supply that salt is now being imported from a number of overseas sources.

For maximum benefit, salt must be applied in anticipation of icy conditions. This means that it is necessary to attempt to predict the occurrence of these conditions. Many motorists have experienced both the difficulties of driving on icy roads and, on the other hand, the frustration as ratepayers of seeing salt spread on roads on which anticipated icy conditions did not materialise. In the first case, delays caused by road blockage and the necessity to drive with greater than normal caution cost a great deal of money to which must be added the cost of any resulting accidents. In the second case, while any one salting foray is probably not too expensive compared only with the cost of delays, the cumulative cost—about £7 million for salt alone—is high. To this must be added the cost of enhanced rates of corrosion of motor vehicles. Estimates of this are now being attempted (Bishop, 1969) and show that corrosion costs about £50 to £60 million per year. This large sum may be reduced by a combination of two methods, the use of additives, which are being developed to inhibit the corrosive effect of salt, and improved weather

155

forecasting techniques to reduce the amount of salt being spread. This chapter is concerned only with the latter.

Current Methods for the Provision of Warning of Icy Roads

Meteorological Office warnings
This is the usual method at present. At Meteorological Offices responsible for the issue of warnings to local highway authorities, the forecaster first estimates the minimum air temperature likely to be experienced in his area during the ensuing night, using well-established techniques. He then estimates the probable minimum temperature on roads and if this is 0°C or below, he issues a warning to this effect, indicating also the time by which road-surface temperatures are expected to fall below freezing point and the probable duration of this condition. Highway authorities prefer to receive these warnings during the afternoon so that, if the roads are also expected to be wet and ice thus likely to form, arrangements can be made for the salting crews to be on call at a suitable time. Around midday then, the forecaster must begin to weigh up the pros and cons of the synoptic situation. One disadvantage of such a time schedule is that, as with all forecasts, the further ahead the warning is issued the more likely it is to be in error. Another disadvantage is that there are occasions when some unexpected development calls for the issue of a warning, or for the cancellation of a warning issued earlier, outside normal working hours. The economic and staffing problems confronting some authorities in acting on warnings or amendments outside normal working hours do not apply to motorways, however, as staff are normally always on duty at the maintenance depot during the winter months and so can be called on at short notice. One of the difficulties facing the forecaster is that the information available to him falls far short of what he would really like. The thermal characteristics and hence the temperatures of the roads in his area might vary appreciably from place to place but meteorological observations are made only at widely separated and well-exposed points, supplemented by less detailed observations from other sources such as motorway compounds, police and fire stations, and AA town offices.

Another point is that temperature measurements refer to air at a standard height of 1·25 m. The relationship between minimum air and road surface temperatures has been discussed recently in a number of papers (Hay, 1969; Parrey, 1969; Clark, 1969; Parrey, Ritchie and Virgo, 1970; Ritchie, 1969) and it appears that while the differences are generally small there are occasions when they can be as much as 3°C or so.

Another problem concerns frost formation. On motorways and most main roads, the flow of traffic during the night is normally sufficient to inhibit frost formation. Frost is thought not to be a general hazard on motorways and trunk roads, but minor roads may be subject to heavy deposits. It appears that the adhesion of a tyre on a frosted surface is reduced only after the passage of traffic has compacted the frost. Alternatively, when the surface temperature is around 0°C, the frost may melt with the passage of vehicles and the water thus formed subsequently freeze.

Ice warning devices

Devices are available for installation in roads to give a prediction of the formation of ice on the road. These, however, are not in wide use and their development is still proceeding. Basically, the device consists of a thermometer adjacent to a pair of electrodes, all embedded in the road surface. The electrodes sense moisture on the road surface through an increase in conductivity between them and, in theory, when the road temperature drops to a certain level ice formation can be expected if the road is moist. There are two disadvantages in this scheme. First, the road surface temperature at which the warning is to be given (assuming a wet road) has to be set to comparatively high levels to give salting crews sufficient time to act before ice actually forms. This 'time factor' disadvantage is, of course, common to all predicting devices. The second disadvantage is that, owing to the hygroscopic nature of common salt, a dry road that is still affected from a past salting can show much the same electrical conductivity as a wet road without salt. It is obviously important to be able to detect the presence of salt, be the road wet or dry, in order to reduce any successive saltings and thus to prevent excessive spreading.

There is no practical method of determining precisely the amount of salt in solution on the road. To be able to do so would seem to be highly desirable in order to limit future salt spreadings and yet obtain an optimum salt concentration. However, existing salt-spreading methods cannot be regulated

G

to such a degree of precision and so approximate methods of 'inferring' the presence of salt have been developed to limit unnecessary applications but not to eliminate them completely.

To provide the inference of salt, two additional sets of electrodes are used. One set is embedded in the road adjacent to the original moisture sensors and a second is embedded in a block which is placed close to the carriageway but sufficiently far away to be unaffected by salting operations. The sensors in the road are designed to indicate levels of conductivity corresponding to a wet road and a wet salty road while the 'off-carriageway' sensor indicates a wet road. By suitably interconnecting the sensors it is possible to deduce that the road is either wet or dry and/or salty and thus give a suitable alarm if the surface temperature drops to a given level.

The presence of snow can also be deduced from yet another set of moisture sensing electrodes which operate in conjunction with a small heater designed to melt the snow and give a moisture indication. Usually snow is sufficiently wet to count as moisture but on occasions both road and snow are dry enough to require the snow to be melted for detection. Fig. 13.1 shows how all the sensors are interconnected to give a warning to spread salt. A logical representation of the system is given in Fig 13.2.

The aim of salting operations is prevention, and warnings must be given in sufficient time for the roads to be treated. In any forecast system there is an overall probability of success or failure and in this context there is the additional probability that the occurrence will happen within a given time. If it takes a time, T, to complete the salting then a freeze within the period T is a 'partial' failure because part of the road remains unsalted. Table 13.1 shows the amount of warning and number of false alarms given by various settings of the 'alarm' temperature of an ice warning device. The table indicates that while the probability of a false alarm decreases with a lowering of the temperature of the alarm setting, the probability of being unable to complete the salting in time increases sharply. If one makes the assumption that the salting proceeds at a uniform rate and that an accident due solely to ice (ie, it would not have occurred if there had been no ice) can occur anywhere over the length in question, it is possible to calculate that for the three alarm settings in Table 13.1 the *relative* probabilities of an accident, due to salting not being completed, are 1, 2·7, and 5·2. It is almost impossible to assess the

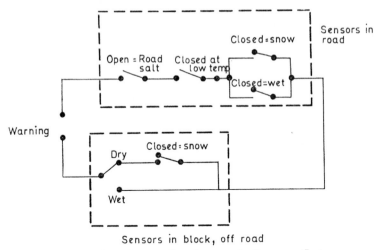

Fig. 13.1   Interconnection of sensors to give warning to spread salt

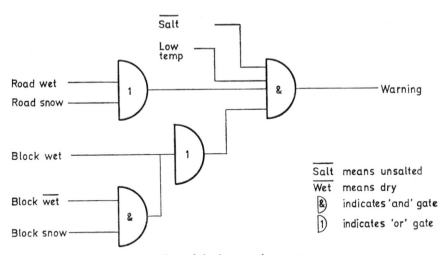

Fig. 13.2   Logical representation of the ice warning system

true probabilities as conditions vary so much, but on the basis of a subjective assessment 1·5°C is usually chosen as the alarm setting.

Table 13.1

PERCENTAGE FREQUENCY OF GIVEN WARNING PERIODS AND
FALSE ALARMS FOR DIFFERENT TEMPERATURE SETTINGS OF
AN ICE WARNING DEVICE

Alarm temperature settings, °C	*Frequency recorded, per cent*						
	Amount of warning, hours						
	$< \frac{1}{2}$	$\frac{1}{2}$–1	1–1$\frac{1}{2}$	1$\frac{1}{2}$–2	2–5	> 5	*False alarms*
1·5	0	10	9	17	27	6	31
1·0	4	23	17	10	18	2	26
0·5	25	41	13	4	6	0	11

Limitations

It must be admitted that the existing ice warning systems leave much to be desired. Tests already carried out and referred to below suggest that a good highway authority foreman can use his local knowledge to improve upon the present ice warning systems and the icy-road warnings as formerly issued by the Meteorological Office. The value of local experience has in fact been recognised by the decision of the Meteorological Office, following consultations with users of their warning service, to suspend warnings of icy roads in favour of warnings of road temperatures below freezing point.

Over a period of two winters, for a road in a rural area there were approximately a hundred Meteorological Office warnings, fifty-seven warnings from an ice warning device and forty-six actual saltings. During another trial covering two motorways for two years each, there were thirty-five warnings from a device and forty-two salting operations. Observations during the latter periods showed that there were only thirteen occasions when some evidence of ice was noted, on nearby roads. These figures show that there is scope for improvement. It is felt, however, that the evidence of these trials should not be given too much weight as they were not properly controlled in that the local maintenance gangs exercised, as it were, right of veto over the automatic device and spread, possibly, unnecessary salt. This then interfered with the operation of the instruments, making it difficult to infer what actually happened.

Many local authorities are not sufficiently well organised to carry out salting operations at short notice, especially at night-time. This has led to the practice of some authorities demanding warnings during the early afternoon, as already noted, for possible action later on. In these situations maintenance gangs feel obliged, as they cannot gamble with public safety, to make precautionary saltings during the late afternoon even though an automatic device in the road says that saltings are not yet necessary. Unless local authorities are willing to improve their winter maintenance procedures there cannot be a significant improvement in salt warning systems, however sophisticated the technology used in them. In the recently published *Report of the Committee on Highway Maintenance* (HMSO, 1970), a recommended procedure for clearing snow and ice is given which if adopted will make the warning system proposed in the next section viable.

A Proposed Warning System for the Motorway Network

It can be seen that each of the existing warning systems possesses defects. The human forecaster knows what the future weather is likely to be but has no knowledge of present road surface conditions, while the ice warning device can record what these conditions are but has no record of future weather. A combination of the two systems, however, would appear to have many advantages. At present the problem of relaying the data back to a forecasting office is too great but will eventually be simplified on motorways, as the whole motorway network will have a communication system controlled by six computers. This system will be able to interrogate meteorological sensors and it is proposed that it should do so and then send the data via a suitable link to local meteorological forecasting offices.

The proposed communication system will consist of:

(a)   voice frequency cable links throughout the motorway network;

(b)   six computer centres;

(c)   thirty control offices;

(d)   one thousand (approximately) outstations in the motorway network, each outstation to include at least one traffic signal.

The data-transmission part of the system which links control offices, computer centres, and outstations is at present operating on the M4 in Gloucestershire and in the Metropolitan Police Area and on the M1, the A1 (M), and the M18 in Yorkshire.

At the moment, the existing computers control traffic signals only, but the whole communication system is capable of dealing with traffic control, the acquisition and processing of traffic and weather data, the control and identification of the emergency telephone network, centralised traffic counting, and the monitoring and fault reporting of the communication system itself. The system is of a modular design to allow for the addition of other facilities such as oral communication with drivers, 'in-cab' visual signals, automatic operation of traffic signals, remote control of vehicles, and communication between motorway maintenance depots. Each computer centre will have a pair of central processors, to allow for the possibility of breakdown, each processor having a minimum storage capacity of 32,000 words and an access time of 1 μs. Each computer centre can serve up to eight control offices which will be situated in general in Police Force headquarters. The control offices comprise a typewriter keyboard, an electronic display, and a 'mimic' diagram of the motorway. The keyboard and display are for the dialogue between the operator and computer. The mimic diagram shows which signals, if any, are in use.

In initial trials with meteorological sensors, to be carried out on the M62 Trans-Pennine Motorway and on the M4 London to South Wales Motorway, the following will be installed and connected to the communication system:

(a)   the basic 'ice warning' package as previously described;

(b)   dry and wet bulb thermometers at the standard height of 1·25 m;

(c)   a rainfall recorder;

(d)   an anemometer and wind-direction vane.

Not all will be connected at the same point, as ice and snow hazards do not necessarily coincide with possible wind hazards. The rainfall recorder will be used in addition to the moisture sensors in the road for road-wetness assess-

ments. The air temperature and wet bulb depression indicated by the thermometers will be used for possible frost predictions although, as already
stated, this is not thought to be a particular hazard.

The various sensors will be connected to a 'black box' containing the A/D
converter and a circuit to produce a digital representation of the items being

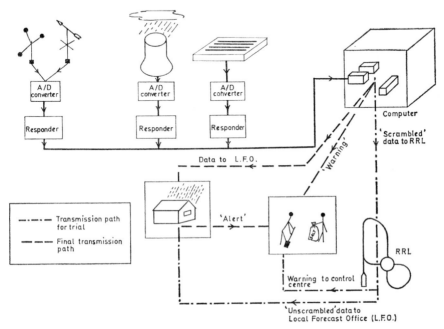

Fig. 13.3 General arrangement of the trial motorway weather forecasting system

monitored. The black box will be connected to a 'responder', one of the
thousand outstations, which will be interrogated every five minutes by a
computer. On interrogation the responder will transmit back to the computer
in digital form the current values of the data being monitored. The data will
then be transmitted via the Post Office 'Datel 200' system to a satellite computer at the Road Research Laboratory where they will be 'unscrambled'. The

satellite computer will assess the data and, if various criteria such as temperature levels are met, will transmit the data in readable form to the appropriate Meteorological Office using the Post Office 'Datel 100' (Telex) system. Figure 13.3 shows the general arrangement of the system.

In the trials, the satellite computer at the Laboratory rather than the main motorway computers will be used to assess the data. This will make it easier to monitor progress and to change the program if experience shows this to be necessary. The programs in the main computers will have many tasks to carry out, such as signal operation and traffic surveillance, and cannot be altered readily.

The intention initially is that forecasters will not actually use the motorway observations in deciding on warnings but will merely note whether or not they would have been helpful. The computer, being supplied with information about the road conditions, can make the logical decisions necessary to simulate warnings produced by the device described on pp 157–60, and these warnings will be passed to the control offices. Later, the various warnings will be compared with actual salting expeditions. Forecasters will no doubt use the extra data eventually if experience shows that this would lead to improved forecasts.

It can be seen that the system described gathers data from motorways only and so, strictly speaking, warnings will apply only to them. However, a motorway is a sample of the roads in a given locality so it is not unreasonable to suppose that forecasts based on motorway conditions will have wider application. It must be remembered that at present forecasts are issued without up-to-the minute information about conditions on any roads.

Acknowledgement

This paper is contributed by permission of the Director of Road Research. Crown copyright.

References

Anon (1970). *Report of the Committee on Highway Maintenance.* HMSO, London.
Bishop, R. R. (1969). *Corrosion of motor vehicles by de-icing salt.* RRL Report No 232, Ministry of Transport, Road Research Laboratory, Crowthorne.

CLARK, C. M. (1969). *Forecasting the night minimum temperature of a concrete surface in winter.* Forecasting Techniques Memorandum No 17, Meteorological Office.

HAY, J. S. (1969). Some observations of night minimum road temperatures. *Met. Mag.*, **98**, 55–9.

PARREY, G. E. (1969). Minimum road temperatures. *Met. Mag.*, **98**, 286–90.

PARREY, G. E., RITCHIE, W. G., and VIRGO, S. E. (1970). Comparison of methods of forecasting night minimum temperatures on concrete road surfaces. *Met. Mag.*, **99**, 349–55.

RITCHIE, W. G. (1969). Night minimum temperatures at or near various surfaces. *Met. Mag.*, **98**, 297–304.

CHAPTER FOURTEEN
R. C. GOODHEW
and E. JACKSON

Weather Forecasting and River Management

Introduction

The forecasting, monitoring, and control of water movement forms an essential part of the proper management of a river basin. The implementation of effective water control strategies, both short- and long-term, compels increasing attention to all aspects of the hydrological cycle—both natural and artificial. One of the water industry's major requirements is the correct forecasting in time and space of the extremes of water availability, ie, heavy rainfall, rapid snowmelt, and droughts. In particular, quantitative estimates of rainfall volumes and maximum intensities are required for relatively small catchment areas. Temperature and evaporation forecasts would also assist in the assessment of the state of a catchment.

This paper considers the dependence of hydrological forecasting upon reliable weather forecasting, and the importance of improving these forecasts in order to minimise the cost of water supplies and to reduce flood damage. The contributions of radar, telemetry, and catchment models are considered in the light of current weather forecasting developments.

Water Conservation

General

The increasing demand for water requires the provision of artificial storage works to even out the extreme variability in space and time of precipitation. Storage units are primarily based on natural catchments, but the urgent need to utilise fully all existing available storage—in order to minimise sociological disturbance and increases in the cost of water—has led the water industry to consider resources on a regional rather than a local scale. This involves the

166

possibility of pumping water from existing storage sites across watersheds to regulate rivers in adjacent catchments. The provision of 'tailored' weather forecasts for schemes such as these will require forecasters to take full cognisance of the provinces which ultimately influence river flows in a catchment.

River Authorities have the duty of managing and making available water resources in England and Wales as advised by the Water Resources Board. They will be increasingly concerned with, amongst other things, the day-to-day operation of regulating reservoirs and the transfer of water so as to provide the maximum benefit and minimum inconvenience to all river users. Considerations include: water supply, agriculture, industry, power generation, navigation, land drainage, flood alleviation, effluent dilution, fisheries, and amenity.

On regulated rivers the reconciliation of the conflicting requirements of river users demands the development of sensitive day-to-day operating rules for the regulating reservoirs. These rules must be amenable to rapid weather changes and consequently are critically dependent for application upon the receipt of accurate and frequently up-dated weather forecasts. Up-to-date indications of river and catchment state are also required in order to make optimum operational decisions.

Long-Term Forecasts

The design of regulating reservoirs depends, *inter alia*, upon assessments of the yield of the reservoired catchment and of the frequency of river flows at salient points. New surface reservoirs may take up to ten years from conception to commissioning, so the significance to be placed upon extrapolation of current estimates of climatic change when considering design criteria must not be undervalued.

Annual and seasonal climatic forecasts are useful at the design stage of regulating reservoirs in order to determine flexible, gross operating rules which may then be interpreted according to the reliability of weather forecasts in the future. These rules ensure that maximum flood retention storage is provided during the winter flood season without seriously restricting hydro-electric power generating capacity at times of peak demand, but also that adequate stored water is obtained for release to the river in summer to

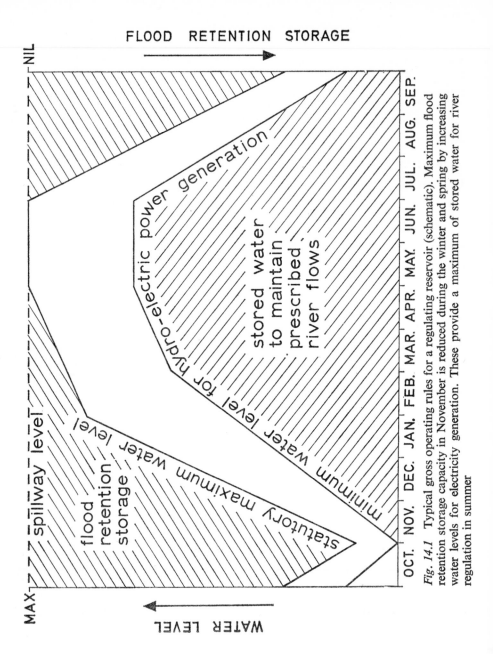

Fig. 14.1 Typical gross operating rules for a regulating reservoir (schematic). Maximum flood retention storage capacity in November is reduced during the winter and spring by increasing water levels for electricity generation. These provide a maximum of stored water for river regulation in summer

maintain prescribed flows during drought conditions (Fig 14.1). Monthly to fortnightly forecasts of precipitation volumes are necessary to predict overall release requirements in autumn, since large regulating reservoirs cannot be rapidly drawn down to provide adequate flood retention storage with only twenty-four hours' warning of heavy rainfall without causing 'artificial' flooding downstream.

Medium-Term Forecasts
Monthly to weekly forecasts of drought conditions during the growing season are a guide to the quantities of spray irrigation water which are abstracted for agricultural purposes from streams. This water, which is largely evaporated, is increasingly obtained from farm reservoirs in order to forestall abstraction restrictions which are currently imposed on unregulated rivers during conditions of very low flow.

The variability of rainfall is so large in comparison with other hydrological parameters that, until quantitative precipitation forecasts are markedly improved, the usefulness of forecasting evaporation rates is limited. However, the accurate forecasting of effective rainfall would be useful in assessing the development of soil moisture deficits and hence in estimating spray irrigation water requirements and recharge to underground storage. In turn, a knowledge of underground storage volumes assists in the prediction of hydrograph recession rates and hence in regulation release requirements.

Short-Term Forecasts
Weekly and daily weather forecasts assume paramount importance when day-to-day decisions are required concerning control operation on reservoirs or rivers. The accurate provision of detailed rainfall forecasts would greatly increase the efficiency of a regulated river system by improving the forecasting of tributary flows into the river between the reservoir and control point (Fig 14.2). The control point flow determines the extent of releases from the reservoir according to prevailing operating rules, but frequently it rains over the tributaries immediately subsequent to a release—resulting in wastage of stored water. Telemetry of river and reservoir levels and rainfall are reducing this wastage (see below), but the over-design of future regulating reservoirs would be markedly lessened by the provision of accurate weather forecasts.

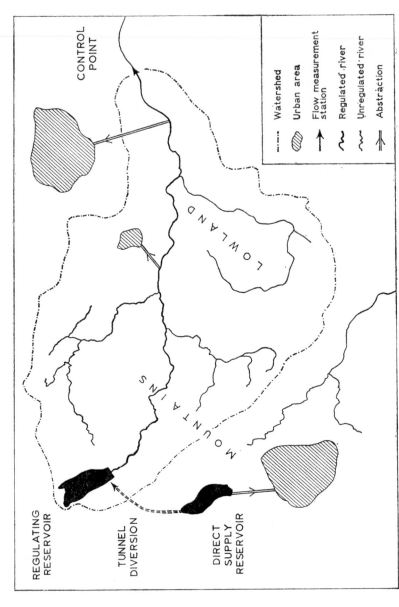

Fig. 14.2 A regulated river system (schematic). The flexibility of the system is increased by diverting water from another reservoir. Releases from the regulating reservoir may take up to five days to reach control point

Flood Forecasting

General

The forecasting of river flooding is intimately dependent upon weather forecasting.

Five definitions are relevant:

(a)	Flood Forecast:	an estimate of the magnitude and time of occurrence of a flood
(b)	Flood Prediction:	the estimation of frequency of occurrence of floods
(c)	Flood Warning:	an indication of the likelihood of floods occurring
(d)	Meteorological Forecast:	a regular information service of expected events
(e)	Meteorological Warning:	a warning of extreme weather conditions which—in the present context—could result in floods

The responsibility for flood forecasting in England and Wales lies with River Authorities who liaise with the police and local authorities for flood warning purposes. Contingency planning and sophisticated instrumentation provide a more reliable information and warning service than formerly, but the weather forecaster is usually the first to be aware of potentially flood-producing weather conditions.

Meteorological warnings of moderate or heavy rainfall and/or impending rapid snowmelt are passed to River Authorities by local Meteorological Offices. The criteria for the issue of warnings are arrived at by joint consultation and depend largely on local catchment and climatic conditions. On receipt, meteorological warnings are interpreted by River Authority staff who may issue alert or standby warnings to the police and others, according to the interpretation of up-to-date river levels, trends, and the state of the catchment. Forecasts of future river levels are made and the police are issued with a flood warning if a potentially dangerous flood situation develops.

Further warnings or cancellations are issued as necessary. The police have the responsibility for warning the public of impending danger or inconvenience according to their knowledge of flood-prone areas.

The importance of a meteorological warning to a flood forecast generally decreases with increasing length of river, since downstream levels can be forecast from a knowledge of previous upstream levels on larger rivers. On small catchments and brooks—particularly if of steep gradient—flood warnings depend solely on alarms actuated by rainfall gauges or alert inhabitants, since there is insufficient time to prepare a hydrological forecast before the flood arrives.

Meteorological Aspects
(*a*) *Heavy Rainfall Warnings*—In some areas, close liaison between the Meteorological Office and River Authorities enables a rapid exchange of information on the developing weather situation and resulting rainfall. Autographic rainfall records supplement the relatively sparse network of hourly reporting synoptic stations. This information assists the weather forecasters in assessing development and movement of rain-areas.

Heavy rainfall warnings typically take this form: 'The rain now falling in the west of your catchment area will spread eastwards and will continue for six to nine hours. Falls will be mainly moderate, but accumulations may exceed one inch locally over high ground in the west'. Knowledge of standard rainfall intensity nomenclature allows rapid estimates to be made of expected storm rainfall accumulations over large areas. As defined, 'moderate' rainfall spans a range of intensities which is of great hydrological significance, for whilst 0·5 mm/h will rarely produce even minor river floods, 4 mm/h could create serious flooding in prolonged storms.

Unfortunately, current practice in Britain does not attach a quantitative index of confidence to the issued meteorological warning. Such a guide would be invaluable in making decisions regarding water releases from regulating reservoirs and standby labour duties.

The full range of information required by River Authorities from a meteorological warning is presented in Fig 14.3. Whilst this does not assume that the present state of knowledge is adequate to supply all the items indicated, existing warnings expressed in a standardised format may improve

1 Severity code

    (i) normal conditions

    (ii) active conditions and highly unpredictable

    (iii) potentially catastrophic situation

2 Catchments liable to be affected

3 Synoptic situation

    (i) direction and rate of movement

    (ii) probability of development of instability type rainfall

4 Storm rainfall

    (i) duration

    (ii) total fall

    (iii) maximum intensity expected—including time and location

    (iv) time of end of heavy rain

    (v) nature of rainfall; viz: frontal, orographic, showery, thunderstorms, widespread thundery rain, or combinations of these types

5 Confidence code for

    (i) timing of rainfall

    (ii) total fall

    (iii) maximum rainfall intensity

    eg, percentage probability of occurrence, or code to represent:

    A   almost certain to occur

    B   probable

    C   possible

Fig. 14.3   A specification for heavy rainfall warnings

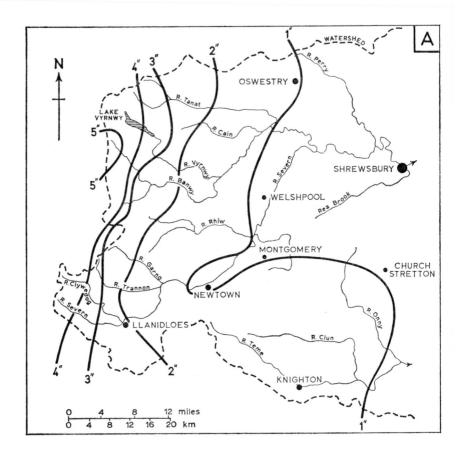

Fig. 14.4 The rainfall which produced the Newtown (Montgomeryshire) flood in 1964

(A) antecedent conditions: 5 to 10 December

(B) storm rainfall: 11 to 12 December (*shown opposite*)

Note the high rainfall gradient in the western, mountainous part of the catchment

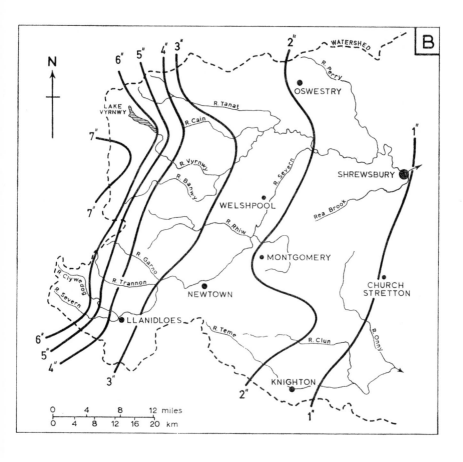

their interpretation in terms of flood-risk by non-meteorological users. Such a presentation would assist in the early identification of areas at risk, such as those susceptible to prolonged orographically dominant storms (Fig 14.4). Furthermore, use of a confidence code should allow as long an outlook as possible to be given.

(b) *Forecasting Research*—Rainfall-producing systems are susceptible to the development of meso-scale features which are often not detected by the national synoptic observing network. These features may take the form of pockets or bands of heavy rainfall embedded in otherwise moderate rain-producing frontal zones. Such conditions can cause serious floods in small catchments if not forecast, since damage costs depend more on maximum flood level than on duration.

A consideration research effort is under way in Britain to improve, *inter alia*, quantitative precipitation forecasting. Numerical modelling by the Meteorological Office of atmosphere fields has already improved the quality of twenty-four-hour general weather forecasts in Britain and the new ten-level model promises to provide more accurate estimates of rainfall intensities in frontal situations. In addition, the Meteorological Office is developing Project Scillonia to provide increasing insight via radar into the meso-structure of rain-producing mechanisms and the relationships between these mechanisms and large-scale synoptic situations. Other research has produced useful empirical rules to improve rainfall forecasts for particular areas on seasonal, monthly, and daily time scales.

The time is thus approaching when it will be possible to forecast general rainfall with some accuracy, but since the predictability of the development of meso-scale and convective systems is often only of the order of a few hours, local forecasters will continue to be required in order to interpret computed forecasts in the light of local conditions.

(c) *Snowmelt Warnings*—During prolonged cold spells, the quantity of water retained in snowpacks gives cause for concern to river engineers. A rapid thaw frequently induces floods because temporarily frozen ground provides an impermeable base which creates considerable surface runoff of melting snow and rain. As a contingency, snow surveyors measure the water equivalent of lying snow in many parts of a catchment as an aid to assessing the likely flood-risk should a rapid thaw ensue. The forecasting of snowmelt rates depends primarily on surface temperature forecasts. The accuracy required is particularly difficult to achieve since advection of shallow air pockets can result in rapid temperature changes at the surface in an otherwise homogeneous airstream.

Hydrological Aspects

(*a*) *Rainfall Measurement*—Regular measurement of rainfall in Britain started in 1860 and has developed over the years until there are now more than 6,000 rain gauges read daily, or in some case monthly, mainly by voluntary observers. From the primary data, various types of information are derived, such as long-term averages, frequency analyses, and extremes, which are used for design work on drainage, water conservation, and flood alleviation schemes. On the larger river systems, daily rainfall data have been essential as a basis for flood warning and have also been a major factor in the day-to-day control of river regulation schemes. The requirement for information on short, intense storms, which have a major runoff effect on small catchments, led to the development of the recording or autographic rain gauge which in its most modern form uses cassette-loaded magnetic tape to record up to three months' data (based on readings taken at fifteen-minute intervals). The magnetic tape is subsequently translated onto punched paper type and may then be used for analytical purposes.

Coincident with this development is the interrogatory gauge which can pass information by Post Office telephone on demand or may actuate an automatic remote alarm system.

Installation of about sixty magnetic-tape recording rain gauges in the upper Dee catchment for research purposes has coincided with investigations into the direct measurement of rainfall by radar. It is hoped to overcome the difficulty with the present system of point measurement which, particularly in mountainous areas, may not give representative values of areal rainfall. The equipment being used is an S-Band (10 cm) radar installed on a hill top at 400 m above sea level. The 4 m diameter dish aerial is capable of continuous rotation in the horizontal plane and a nodding motion in the vertical. Radar signals are integrated, digitised, and recorded on magnetic tape together with rainfall data from three calibration areas. These areas, of approximately $2\frac{1}{2}$ km², are between 13 and 32 km from the radar and each contains clusters of five well-spaced, tipping-bucket rain gauges connected to a radio-link telemetry system.

Radar measures a quantity Z (the radar reflectivity factor) some distance above the ground. This can be related to the surface rainfall rate R by $Z = A R^{1 \cdot 6}$, where A depends on factors such as the radar system calibration, the rain drop-size distribution and changes in rainfall rate below the beam. The clusters of rain gauges enable A to be determined and thence R can be obtained. The horizontal range of the radar for rainfall measurement is 50 km, but an assessment of incoming rain in terms of direction and speed can be obtained from PPI (Plan Position Indicator) representation up to 400 km.

It would seem that ultimately a number of rainfall-measuring radars suitably interspaced over the country may meet both the needs of the Meteorological Office and the River Authorities. They could replace many existing rain gauges and, it is hoped, give improved results. The initial capital costs and running costs although high would probably be only a small percentage of the savings in potential damage costs made by the provision of adequate flood warnings.

(b) *River Measurement and Telemetry*—Daily climatological and hydrological data are used primarily for non-repetitive design studies but a continuously available supply of information is required for forecasting and river management. For some years the 'Telytone' type of instrument has been used for transmitting information on water levels over the Post Office public subscriber network, and more recently interrogatory rain gauges have reached a state of reasonable reliability. In addition, devices at outstations can now give alarms by initiating calls over the Post Office network to a central control. These instruments have been used by River Authorities to provide flood warning systems of varying degrees of complexity and efficiency, and constitute a simple form of telemetry (the remote reading of measuring instruments).

The advent of river regulation schemes, mainly initiated for the purposes of water supply, has shown the need for sophisticated operational procedure which, by the use of telemetry, would ensure that best use is made of sluices, flood storage basins, reservoirs, and pumping stations, both on the grounds of effective use of capital and of better flood warning. Such a system requires a continuing scan of field outstations from a central control enabling more accurate forecasts to be made and regularly updated.

A modern system recently installed within the River Dee catchment may be described as follows:

A small computer arranges the automatic scanning of outstations at preset intervals (five, ten, fifteen, thirty, sixty minutes and three hours, or on demand). The links to the outstations are via an interface and then one of three transmission methods:

(a) UHF radio,

(b) PO private line,

(c) local private lines.

Data on rainfall, river levels and lake levels are returned to the computer in digital form and after processing are presented in the control room in four modes:

(a) Teletype prints out river levels, lake levels, and accumulated rainfall— continuously and daily from 09.00 hours—alarm conditions and failures, date, and time.

(b) Punched paper tape recorder collects basic data which will be recorded at half-hourly intervals independently of the scan rate used on the teletype.

(c) Mimic diagram indicates levels and alarms at a number of key points and gives direct visual assistance to the operator.

(d) Multipoint trend recorder indicates graphically the change of state at up to twelve outstations with a coverage of twelve hours visible on the chart.

As the telemetry scheme is proved and developed it is anticipated that the computer with its storage facility will be used to provide a teletype output of calculated river discharge (flows), lake volumes, and incremental (half-hourly) values of rainfall. Time-lagged discharges involve storage of certain back data for up to a maximum of two days.

This system is expensive both in capital investment and in maintenance costs. The justification for it in this particular case is based on the increase in yield for water supply which can be achieved by more refined river regulation. However, one can expect more prompt, realistic flood warnings and more accurate flood forecasts as additional benefits.

Between this sophisticated system and the simple interrogatory units there is obviously room for other types, and the Post Office has a number of developments in hand. One such is the Data Control Equipment 1A. This provides for the automatic, sequential dial-out over the public telephone system to terminals which give an automatic answer. In conjunction with suitable modems it 'dials' the call, obtains an acceptance signal from the distant station, then puts the circuit into a state of readiness for the transmission of data. Should the call be ineffective the equipment will abandon it and transfer to the next call. Arrangements can be made for the presentation of the information on a teletype instrument.

Thus it is evident that in many areas which require warnings or forecasts of floods, the simpler and cheaper systems are likely to be used for some time in preference to the accurate but more expensive ones required for river regulation for water supply. A twenty-four hour watch is necessary with equipment operating continuously and requiring expensive maintenance.

(c) *Hydrological Forecasting and Catchment Models*—A knowledge of available storage in soil and surface depressions is essential if river forecasting is to be attempted from rainfall information. The storages develop when evaporation exceeds rainfall. The resultant soil moisture deficits are a guide to the proportion of rainfall which may be expected to form surface runoff and create potential river flooding.

The operational application of soil moisture deficits to river forecasting is a recent procedure which has awaited the development of automatic climatological stations to provide up-to-date data via telemetry for the estimation of evaporation. Installation of this equipment is now proceeding, so improved river forecasts may be anticipated.

The 'moistness' of a catchment may be less rigorously represented by an antecedent precipitation index. This index weights rainfall according to its contribution to current ground moisture conditions and, by correlation with past surface runoff volumes, provides a simple guide to potential flood-risk.

The distribution of surface runoff in time is effected by use of the unit hydrograph for a particular river gauging station. Subsequent forecasting of downstream river levels on the larger rivers proceeds by routing the flood hydrograph through channel and flood-plain storage.

In the past decade, attempts have been made in Britain to use some or all of the above factors in assessing catchment response, particularly in the flood forecasting field. However, the successful use of the hydrograph and routing methods requires precise data, particularly of rainfall and river flows, and hitherto these have not been available. A more recent approach attempts to reproduce mathematically the action of a catchment and a river system. This mathematical model should satisfy the following conditions:

(a) It should represent the significant hydrological processes, and

(b) contain the minimum number of parameters for their representation.

(c) The parameter values should be measurable.

(d) Up-dating procedures should be incorporated.

On a programme of research currently being undertaken on the River Dee the behaviour of the catchment and a number of reservoirs is being simulated on a digital computer. Then, using a long series of flow data as input, the best way of operating the reservoirs can be established in respect of floods and droughts. The next step involves a model representing seven sub-catchments into which will be fed hydrological data from the telemetry system on a permanent operational basis. The output, incorporating the correct setting of reservoir outlet controls, will minimise any wastage of water and the effects of flooding.

As a model becomes more complex in an effort to represent the catchment more fully, the possibility of input-data error becomes increasingly significant. Whilst this problem can perhaps be easily dealt with on research projects within a university laboratory, it becomes critical if the model is put to operational use. Thus an incorrect input could create a chaotic situation whereby the overall assessment could be lost and a control operator would possibly be worse off than if the model were not available. Hence, there is every inducement to make the model as simple as possible so that it can be easily readjusted to zero. An automatic, parameter-optimising routine has recently been developed to speed up the 'locking-on' of a model on to the particular catchment concerned.

Conclusion

The refinement of meteorological and hydrological forecasting procedure is likely to lead to long-term savings in the cost of potable and industrial water on regulated rivers. Although regulation will be increasingly used in the future, current emphasis in hydrological forecasting is mainly directed towards savings on flood damage to property and crops and industrial losses.

The wide use of weather forecasts enables the Meteorological Office to justify considerable expenditure on research and development. At the same time the growing demand for water will mean greater research by the Water Resources Board and capital investment by the River Authorities in instrumentation for hydrological measurements.

The capital costs involved in the sophisticated equipment with its high running and maintenance costs must be weighed against the benefits obtained. The increasing need for local weather forecasting offices to receive accurate, local, up-dated weather observations at a time when River Authorities are installing telemetered, automatic climatological and rainfall stations for water-management purposes, would indicate that some rationalisation of local meteorological and hydrological data facilities in Britain is desirable.

Acknowledgement

The authors wish to record their thanks to their respective Authorities for permission to publish the material in this chapter. The views expressed are those of the authors and not necessarily those of their Authorities.

Bibliography

BENWELL, G. R. R. (1967). The jet stream at 500 mb as a predictor of heavy rain. Met. Mag., 96, 4–9.

BENWELL, G. R. R. and BUSHBY, F. H. (1970). A case study of frontal behaviour using a 10-level primitive equation model. Quart. J. Roy. Met. Soc., 96, 287–96.

BLEASDALE, A. (1970). The rainfall of 14th and 15th September 1968 in comparison with previous exceptional rainfall in the United Kingdom. J. Inst. Wat. Engnrs., 24, 181–9.

BROWNING, K. A. and HARROLD, T. W. (1969). Air motion and precipitation growth in a wave depression. Quart. J. Roy. Met. Soc., 95, 288–309.

BROWNING, K. A. and HARROLD, T. W. (1970). Air motion and precipitation growth at a cold front. Quart. J. Roy. Met. Soc., 96, 369–89.

BUSHBY, F. H. and TIMPSON, M. S. (1967). A 10-level atmospheric model and frontal rain. *Quart. J. Roy. Met. Soc.*, **93**, 1–17.

BUSSELL, R. B. and JACKSON, E. (1968). Telemetry for River Authorities. *J. Inst. Wat. Engnrs.*, **22**, 196–201.

COLLINGE, V. K. and CRANN, H. H. (1970). The Dee research programme. *Proc. Inst. Civ. Engnrs.*, **47**, 549–51.

GOODHEW, R. C. (1970). Weather is my business, I. The hydrologist. *Weather*, **25**, 33–9.

HARROLD, T. W. (1965). Estimation of rainfall using radar—a critical review. *Met. Office Sci. pap.*, **21**.

IBBITT, R. P. (1970). *Systematic parameter fitting for conceptual models of catchment hydrology.* PhD thesis, University of London.

JOHNSON, P. (1966). Flooding from snowmelt. *Civ. Engrg. Pub. Wks. Rev.*, **61**, 747–50.

LLOYD, J. G. (1968). River Authorities and their work. *J. Inst. Wat. Engnrs.*, **22**, 343–402.

LOWNDES, C. A. S. (1968). Forecasting large 24-hour rainfall totals in the Dee and Clwyd River Authority area from September to February. *Met. Mag.*, **97**, 226–35.

LOWNDES, C. A. S. (1969). Forecasting large 24-hour rainfall totals in the Dee and Clwyd River Authority area from March to August. *Met. Mag.*, **98**, 325–40.

MASON, B. J. (1969). Some outstanding problems in cloud physics—the interaction of microphysical and dynamical processes. *Quart. J. Roy. Met. Soc.*, **95**, 449–85.

MASON, B. J. (1970). Future developments in meteorology: an outlook to the year 2000. *Quart. J. Roy. Met. Soc.*, **96**, 349–68.

MURRAY, R. (1968). Some predictive relationships concerning seasonal rainfall over England and Wales and seasonal temperature in central England. *Met. Mag.*, **97**, 303–10.

MURRAY, R. (1969). Prediction of monthly rainfall over England and Wales from 15-day and monthly mean troughs at 500 mb. *Met. Mag.*, **98**, 141–4.

RATCLIFFE, R. A. S. (1968). Forecasting monthly rainfall for England and Wales. *Met. Mag.*, **97**, 258–70.

The Demand for Beer as a Function of Weather in the British Isles

This investigation considers the demand on the manufacturer (a group of breweries) made by retail beer purchasers (a conglomerate of beer drinkers) for beer in a variety of qualities, according to weather variation (a local phenomenon), *ceteris paribus*. The work is an applied business problem using historical data economically available and analysed by theoretical econometric techniques of proven validity. Its scope lies within one diversified brewery company with 10 per cent of the UK total annual beer market of 33 million barrels (each of 36 imperial gallons = 163·8 litres), trading through several different kinds of market, over a time series of many years (Monopolies Commission, 1969). A pilot inquiry has been made in respect of a semicircular land area with its centre at Southampton and a radius to more than cover Hampshire. This particular market buys through public houses tied to the company by ownership. The study is concerned with weekly wholesale deliveries of draught bitter beer and uses Southampton Meteorological Office's, daily measurements of maximum temperature, total sunshine, and total precipitation, for the sixteen years starting 1 January 1953. The land area was selected because it had the longest available history of beer data based on unaltered physical boundaries, and minimal changes of other factors affecting beer data such as changing ownership of licensed premises.

Objective
The initial intention has been to establish a weather function for each of the fifty-two weeks of the year and then to use it in two ways: firstly, as a historical correction of the observed beer sales, to provide marketing management

with results clear of the uncontrollable factor of weather effect, and secondly, as a forecasting correction of short-term predicted sales, to provide production management with figures incorporating the effect of expected weather.

Methods

The forecasting correction requires an efficient method of forecasting of the values, for up to three weeks ahead, of selected meteorological parameters. Such a method had been tested satisfactorily by the author on relevant data some time before the present investigation started. The variables were weekly averages of daily records of maximum temperature, hours of sunshine, inches of precipitation, average wind speed, and average humidity, achieving a three week forecast efficiency shown by coefficients of variability ranging from 10 per cent to 1 per cent. The forecasting method used was exponential smoothing over the four factors of moving average, fifty-two-week seasonal ratio, trends by first and second differences (Winters, 1960), with best weights found by computer iterative search over four consecutive years. It was necessary to transform precipitation values logarithmically because of the frequent occurrence of 'trace' reports (Rimmer, 1970). Wind speed and humidity have been omitted in this investigation to date as some earlier values were not available in punched-card form. It is not considered that they should pass uninvestigated, bearing in mind the interesting result of the Central Electricity Generating Board's analysis showing the high correlation between demand for electricity and wind speed on Mondays (washing days) (Davies, 1958).

Practical Relationships

In relating beer sales to weather parameters their practical everyday relationship must be considered. Draught beer is legally sold daily (except in some Welsh counties on Sundays) through 'on-licensed' public premises during permitted hours, generally 11 am–3 pm and 6 pm–11 pm, according to the conditions imposed by the Petty Sessional Division (modified for Scotland), granting the licence to the licensee at the premises (Licensing Acts). These times are virtually constant throughout the seasons of the year, whereas temporal variations exist for the selected weather parameters: daily maximum temperature, daily total hours of sunshine, and daily total inches of precipitation. It may be mentioned at this point that, in attempting a comparable quantity dimensional measure, 24-hourly minimum temperature was jointly analysed and rejected, as were mean 24-hourly temperature and range of 24-hourly temperature, as not contributing additional significant explanation of the variation in the beer sales due to changes in the weather. Ready availability of inexpensive data, in punched-card form from the Meteorological Office, had some bearing on what weather parameters were considered for use.

A further practical relationship is that, while this beer may be drunk daily in synchronism with weather variation, it is drunk from public-house stock. This stock is normally replenished by a routine weekly delivery, the barrelage volume values of which, summed over hundreds of public houses, constitute the observed beer sales of the analysis. This calls for a *one-week time-lag* in the analysis. Delivery abnormality can occur. Greater distance from the depot in relation to mean lower throughput of the public house, and high demand on peak days, are examples of the causes of longer and shorter time-lags, respectively.

The data were therefore tested with variable weekly time-leads of weather on beer. With the time-series used, one week gave the most significant results. There is scope to examine time-lags of the draught bitter concerned in subclassified categories of 'chilled and filtered' (which here has a shelf-life of well over three months) and 'fined' (with a shelf-life of a week in warm weather) since both products are measured as wholesale sales delivered, —ignoring change in size of stock—and not as retail sales consumed.

Factors in Variation in Beer Sales

A problem arises in separating the observed values of beer sales into those which are deterministically due to known and measured causes and those which are probabilistically due to unknown causes aggregated with weather; it is sought to determine the relationship of the latter to sales. Short-term trend of growth/decay (economic causes? eg, bankruptcy of big local aero-engine works) may be explained by an exponentially weighted moving average (correlated with weather?). Long-term trend (market change? eg, change of taste from mild to bitter beer) may be explained similarly using first and second differences of consecutive observations. Average weather conditions and public holiday peaks and troughs may be explained, again similarly, by a fifty-two-week seasonal ratio, whence a smoothed forecast made in retrospect was able to give expected beer values according to average weather and other known conditions over past years. Recent research has derived more useful adaptive systems (Brown, 1967). It has been established by the author that the beer sales response to own and competitive television advertising stimulation can be evaluated significantly and thence correct allowance made. In this case, for television, the 'noise' in the system did not justify such correction for the elements partially concerned over time, space, and product, thus also not for other unmeasured causes, the totality of all of which were in due course found to constitute half of the 'residual variance' unexplained. In other words weather correction did explain one half.

Public Holiday Peaks
The explanation of public holiday peak/trough demand in relation to changes in the weather is one of the more difficult and important tasks. Beer demand peaks particularly at Christmas, Easter, Whitsun (after 1964, Spring Bank Holiday), and August Bank Holiday (after 1964, moved from the first to the last Monday in August), because of the nature of the holiday. The change in demand as a consequence of weather variation may be greater on these days than on non-public holidays. This effect may be convoluted with the likelihood that at more extreme weather parameter levels, some of which occur at these public holiday times, the relationship is again of a higher order than normally. Superimposed on this is the trade practice of delivery to stock to build up supplies against the forthcoming public holiday—up to a four-week lead. The holiday is followed by an immediate and pronounced trough in demand, usually of one to two weeks' duration, according to the interaction of (a) level of realised peak consumption, (b) lack of post-holiday purchasing power and (c) a delivery week with a Monday which is itself a public holiday from delivering goods. These oscillations are not only chronologically reciprocating for the movable feasts of Easter and Whitsun, but have a similar, though smaller, tendency over Christmas as its constant date may fall on any of seven weekdays while a normal delivery week is of five days. Legal changes in past Bank Holiday dates have already been quoted. It was therefore necessary to standardise public holiday peaks to those of a given

year. It should also be mentioned that as a fifty-two-week year generates more than one fifty-three-week year over sixteen consecutive years this also called for correction. Correction was done by manual adjustment of the relevant data on visual inspection of its graphical representation, but transient oscillations, which might reflect weeks of exceptional weather, were left undisturbed. The errors of such subjective estimates injected noise into the system—clearly identified in due course when an adjusting allowance was sought for it.

Statistical Techniques and Analyses

A decision-function suitable to allocate the effect of concurrent weather might consist of a simple formula relating the appropriate weather parameters to the otherwise expected demand for the product. This can be calculated by the technique of multiple linear regression, when some function of observed and expected beer values is held to be a rectilinear dependant of the mutually independent weather parameters. One part of the problem of the relationship, at more extreme weather levels, is the possibility of curvilinear dependence, particularly around any unknown thresholds of response and saturation, due to weather only. The Gas Council, in a somewhat more capitalised project concerning the demand for North Sea Gas for domestic heating, has used the concept of a degree-day. This assumes nil demand until temperature drops below an arbitrary threshold, and therafter linear temperature demand over time (Badger and Lyness, 1969). The simplicity of this one-sided approach was not overlooked, but it was rejected in favour of the more cybernetic system of self-determining levels. For beer demand the possible sensible transformations of dependent variables were reduced to the difference of *observed* minus *expected* demand, and the ratio of observed divided by expected demand. Against the possibility of skew distributions arising from non-rectilinear causation the logarithms of the differences and the ratios were also considered. It was realised that the chosen transformation would be limited to a sample size of sixteen in a proposed regression on distributions of temperature, sunshine, and precipitation. If all were not normally distributed they could still be symmetrical, and similar, to satisfy the assumptions lying behind the proof of the validity of regression analysis (Plackett, 1962). The four beer value transformations, for each of the fifty-two weeks, sometimes gave poor probabilities of not being due to chance in respect of skewness and kurtosis, when tested for normality. They were found to perform better using *ad hoc* probability values derived from sampling random numbers, sixteen at a time. This led to the selection of the ratio of observed/expected beer sales for the first twenty-five weeks and their log ratio for the next twenty-seven weeks. The latter selection accorded with empirical expectation that in summer weeks the temperature effect might be more powerful than simply rectilinear. An intention to examine the ratio with a denominator using increasingly sized samples of the expected values, up to fifty-two of the year at a time, to reduce error expectations, was kept in abeyance.

The computer regression analysis print-out was arranged to show regression-coefficient significance-probabilities as well as inter-correlations and stepwise evaluations, as variables entered and left the analysis. No real economy in future use of the finally determined decision-function was foreseen from a reduction in the number of weather variables to be used. The stepwise inspection was more a matter of interest because true weather effect on beer sales, while considerable, is complicated. For example, for the London day-trippers' holiday resort of Brighton trade experience repeatedly shows the following ideal weather. London, early morning *sunshine* to induce the hour's train journey to Brighton; Brighton, morning, pub-opening time, *wet* to send them inside; afternoon, pub-closing time, *warm* to persuade them to stay on the beach at Brighton; evening, pub-opening time, *windy* to send them into the pubs again. Analyses of this detail are impracticable because of the cost of the necessary special records of retail sales over the bar, assuming that such special records would be reliable anyway. In any case forecasts of such detailed weather are not available for use as a production managerial aid, even for beers which can be produced in three days with modern methods using the chemical engineering of continuous fermentation.

The fifty-two weekly regressions of beer sales on the three weather variables were examined for the extent to which they explained the variation of the ratio of observed/expected values, by means of the square of the multiple correlation coefficient. While many such R^2 values were high, with associated high probabilities of not being due to chance, the general pattern was disappointingly oscillatory. This pattern remained, though it was reduced, after

H

empirical correction for noise introduced by initial standardisation of the movable feasts. The empirical correction was to increase R^2 by a constant factor of the number of standardising adjustments originally made for each week which had been observed sixteen times. In addition to sources of error already mentioned there is the possibility of transcription errors in our departmental work. Sources of error were also found in the meteorological cards: duplicated dates, missing cards, and mechanically poor cards which could lead to machine reading errors. Of course, these were corrected as they came to light. The false straight line trend over fifty-two weeks' beer sales, caused by the date of the starting point within the seasonal pattern of the year, was eliminated by earlier analysis which had derived a starting week for otherwise trendless values such that no false trend was generated by seasonality. The R^2 values were accordingly chronologically arranged from that starting week. It appeared permissible to apply external general knowledge that adjacent weeks should have similar regression equations of this kind and thence to argue that adjacent oscillations were likely to be due to sampling and injected error. In these circumstances a symmetrical unimodal curve of best fit was empirically drawn through the R^2 values. It gave results that in the lowest winter week weather accounted for 45 per cent of the otherwise undetermined variations in sales; in summer the proportion explained gradually rose to 55 per cent.

The artificial adjustment of R^2, however logical, does not enable one to correct consecutive oscillating equations producing individual weekly polynomial weather explanations. This was done by treating each weather parameter for best fit in the same way as R^2. The measure was the product of the partial regression coefficient and the mean of its variable divided by the mean of the dependent variable. (At this stage it had been decided to forego log ratio for simplicity, in favour of all ratio beer values. The loss of accuracy was of the order of R, dropping from 0·73 to 0·71 for virtually the same level of significance.) The artificial equation for a given week was then reconstructed by the addition of a constant which brought its total value to unity, the mean value of the ratio of observed/expected beer sales. A final measure of the root mean square error function of each of the fifty-two equations is now in hand. There is also in hand a recalculation with arbitrarily truncated public holiday peak/trough adjustments only. It may prove worthwhile to pool public holiday residual values for evaluation against weather, to determine if special coefficients are applicable at such times.

By regressing each week's mean seasonal observed beer sales (that is, detrended, by years, as found in the forecasting model for expected beer sales) against the corresponding means of weather parameters (which are hoped to have no annual trend), fifty-two sets of observations were used to find the coefficients and their significances, in an equation relating percentage change in beer sales to measured weather, common to all weeks of the year. It has already been ascertained that these functions differ between bottled beer and draught beer and in different parts of the British Isles. This accords with the expectation of experience that warmer weather has a different volume consumption effect on a half-pint (more gaseous) bottled-beer drinker than on a full-pint draught-beer drinker, and with the knowledge that draught beer preference increases with latitude in Britain.

Business applications accept lower levels of significance for practical work than academic standards usually require. It has to be determined whether the error size of the fifty-two reconstructed equations, as they run into winter, is such that the common year equation can be used equally effectively in a number of cases.

These common year regressions have the disadvantage of reintroducing the bias of public holiday peaks/troughs in sales. It is a consolation that their temporal locations tend to counterbalance so that most of the bias would appear to go into the constant of the equation. At the same time one has an unhappy memory of analyses of individual weeks where, in a sample size of sixteen, two opposing extreme values have given a false coefficient.

Technical Considerations

Detailed technical considerations led to the sample choice of locality, market, quality of beer, frequency of date, meteorological parameters, length of epoch, and techniques of analysis. Subsequent investigations will be made to repeat the established worthwhile routine over other localities, markets, and qualities of beer, in order to evaluate the differing coefficients in the decision-model to the limit of reliable results. It then becomes a matter of incorporating this in a macro-model extending beyond the weather.

This investigation into the demand for beer as a function of weather has been confined to the almost instantaneous functional relationship between consumption demand and weather. No account has been taken of the fact that consumer reaction to a given weather condition may vary and may be maintained after the weather has changed. It should be stated that the possibility of a long, time-lag, functional relationship between beer quality and weather is not overlooked. It is appreciated that weather growing conditions affect the quality and quantity of cereals, mainly barley for malt, and the hops, which are the principal raw materials of beer. In practice, the brewer is able to smooth the effect of variations in harvest quality and quantity by purchasing from a variety of origins through organisations such as the Home-Grown Cereals Authority and the Hops Marketing Board, which analyse and advise on quality and adjust price on quantity (Britton, 1969). The brewery chemist analyses samples of raw materials delivered and his reports put the brewer in a position of almost complete up-to-the-minute pre-brewing quality control. Even so, there may well be scope to derive econo-weather functions to link likely price with past weather. For instance some 'natural' beers require low-nitrogen barley, and recent harvesting is putting high nitrogen values on the market.

Sources and Acknowledgements
The earlier data of observed beer sales were manually extracted from the ledger records of the management information department of the brewers company concerned, and the later data was printed out from the data banks of its IBM 1440 and 360/40 computers. The weather data was purchased in punched-card form from the Meteorological Office, Bracknell, Berks. The expected beer sales were calculated on IBM's Exfor 1 Program signalled from Newman Street, London, into its IBM 7094 computer at Hursley, Hampshire—though the program is unsuitable for some long time-series and this one had to be processed in four parts.

The calculations were carried out by a poly-program written in Fortran 4G by Mr N. R. Hawkins of the economic and market research department of the brewery company, using an IBM 360/50 computer located at Liverpool and via a landline telegraphic link to Unilever's Terminal at Black-friars, London. Peripheral mathematical statistics were carried out by Mr C. Flack of the same department, using his own Fortran 4G programs on an

ICL 1905E computer at the Northern Polytechnic, London, and he executed cross-checks using the GEIS Terminal at another brewery signalling into a GEG 235 computer.

The National Cost

The values of beer data used and coefficients determined have not been published for reasons of commercial security. Not all big brewery companies have gone to the expense, over the years, of recording weekly data, and few, if any, have incurred the expenditure of this type of research. Therefore, in obtaining the permission of the directors of the brewery company concerned to accept an invitation to write this chapter, the author has not asked that the brewery results should be promulgated in detail. A conservative indication of the potential value of beer-weather correction to the national economy can be obtained. Crude, computerised demand forecasts by product, area and market, week by week, have shown mean errors of the order of 10 per cent. One-half of this unknown is demonstrably explained by weather correction for production purposes (ignoring the intrinsic 10 per cent error found in the short-term weather forecasting for this purpose); it amounts to 5 per cent total volume. The national retail expenditure on beer is over £1,000 million per annum (Blue Book, 1970). With wholesale selling price at 80 per cent of retail selling price (Cmnd 4227, 1969a) and wholesaler's cost at 85 per cent of wholesale selling price (Cmnd 2965, 1966), this 5 per cent total volume is worth £34 million per annum. Regarded as wholesale stock avoidably locking up capital at an average 10 per cent return (Cmnd 4227, 1969b) this wastes *£3·4 million per annum* of the national economy. This is without evaluating the better managerial control on the marketing side. The cost of the complete investigation will be less than one-tenth of one per cent of this wasted sum.

References

BADGER, E. H. M. and LYNESS, F. K. (1969). *A National Survey of the Relative Severity of Past Winters with Particular Reference to Gas Storage Policy.* The Gas Council. Ref. OR 36.

BLUE BOOK (1970). *National Income and Expenditure, 1970.* Central Statistical Office. Table 22, 26.

BRITTON, D. K. (1969). *Cereals in the United Kingdom; Production, Marketing and Utilisation.* Pergamon Press, Oxford.

BROWN, ROBERT G. (1967). *Decision Rules for Inventory Management.* Holt, Rinehart and Winston, Inc, New York.

CMND 2965 (1966). *National Board for Prices and Incomes,* Report No 13, *Costs, Prices & Profits in the Brewing Industry,* 4.

CMND 4227 (1969a), NBPI, Report No 136, Beer Prices, Table IV, 16.

CMND 4227 (1969b), NBPI Report No 136, Beer Prices, Table III, 13.

DAVIES, M. (1958). *The Relationship between Weather and Electricity Demand.* Institution of Electrical Engineers. Monograph No 314S.

LICENSING ACT, 1964. Licensing (Scotland) Act, 1962.

MONOPOLIES COMMISSION (1969). *A Report on the Supply of Beer.* HMSO, London.

PLACKETT, R. L. (1962). *Regression Analysis.* Oxford University Press, Oxford.

RIMMER (1970). *Private communication.* An arbitrary 3-value transform found excellent for rain correlation with ice-cream sales. Unilever.

WINTERS, PETER R. (1960). Forecasting Sales by Exponentially Weighted Moving Averages. *Management Science,* **6,** 324–42.

The Problem of Forecasting the Properties of the Built Environment from the Climatological Properties of the Green-Field Site

The Relevance of Climatological Information to Urban Decision-making

The first question to ask is, 'Why is it important to try to forecast the climatological properties of a built environment in advance of its actual construction?' The answer is, of course, to enable a more effective urban design to be achieved. This then opens the question 'What is a more effective urban design from the climatological point of view?' Only a relatively precise answer to this question will enable one to identify the relevant climatological parameters one needs to forecast. One can then proceed to a consideration of whether it is actually possible to make such forecasts, or not. If not, design must proceed on the assumption that the green-field climate is representative of the built environment.

There are a number of facets of urban design which are very sensitive to climate. We may list some of the important problems.

(a)   Optimisation of land-use patterns in relation to the different activities to be carried out in the town, for example, atmospheric pollution zoning.

(b)   Identification and development of suitable microclimates for various activities, for example, parks, recreation, allotments, school outdoor recreation, swimming, etc.

(c)   Identification of adverse microclimatic factors likely to influence the detailed design of urban systems, for example, high local winds interacting with transportation system design, adverse slopes unsuitable for residential use.

(d)   Optimisation of building form in relation to external climatic inputs. Factors like radiative heat gains and heat losses, ventilation patterns, etc, influence the desirable form and orientation of buildings, as well as the desirable fenestration.

(e)   Optimisation of building form in relation to microclimatic modification of the immediate exterior domain of that building, for example, avoidance of high winds linked with tall buildings in association with unsuitably placed low buildings.

(f)   Constructional safety against high winds, icing, snow loading, etc.

(g)   Selection of building materials for durability and appearance in relation to climate as a dominant factor in deterioration and weathering.

(h)   Planning of construction as an operational process in relation to the hazards, and interruptions presented by the climate.

(i)   Control of water runoff in towns and from buildings; flood control.

(j)   Assessment of probable plant running costs in advance of construction, for example, air conditioning, heat loads, lighting systems, etc.

(k)   Optimisation of the operating environment of transport systems, for example, avoidance of snow drifting, ice hazards.

(l)   Control of the environmental impact of a transport system on its adjacent urban systems, for example, climatological control of vehicle pollution levels to adjacent property.

The Need to Have Information in Advance of Decision-making
In all these situations it is necessary to forecast the probable outcome before construction. The economic penalties for bad climatological decisions are severe, because it becomes impossibly expensive to change an urban system afterwards. It is of no value for the designer to know the properties of the system after construction, because by then it is too late to change it. It follows that climatological model-building involving either theoretical or physical models is an inescapable necessity for improved design, however difficult the practical problems of doing it accurately may be. Urban climatological model-building can be considered as a special sort of forecasting. This forecasting art is relatively undeveloped in Britain, though 80 per cent of the population of the UK lives in towns.

Such model-building clearly has to be statistically based. In urban climatology, special interest centres on extreme value conditions associated with urban failures. Extreme value analysis therefore must always form an important part of the 'art' of urban climatological forecasting.

Building Design Relationship Functions

Actual design requires the availability of suitable relationship functions to relate the meteorological variable to the actual design problem. For example, to find the wind load from the wind velocity, one must know the aerodynamic characteristics of the building, as well as the design wind velocity and associated vertical wind gradients. The wind velocity in isolation is not enough to solve the problem. Thus, following Thom (1970), the urban designer needs to have available suitable relationship functions $f_1, f_2, f_3, \ldots$ to be used in expressions of the general form:

$$t_L = f(t_W) \tag{1}$$

where t_L is the required derived design information, for example, wind load in N/m^2

t_W is the weather design variable, expressed as a random variable with an associated distribution function

t_W may be multidimensional, eg, cloud water drop content, cloud drop size, air temperature, wind speed and direction in the case of ice loading from rime ice. t_W also has vector properties. Fig 16.1 shows a check list which the author has developed to help urban decision-makers identify the multi-dimensional meteorological variables for any particular problem they may have under consideration.

Clearly the urban forecast data must be in a suitable form to be fed into eq (1). Thus it is necessary to have precise knowledge of the different relationship functions to be used in design to identify precisely the relevant meteorological parameters to be forecast for the urban situation.

Derivation of Meteorological Data for Urban Design

Once the first aspect of the problem, namely the identification of the relevant weather variables, has been clarified, it is then necessary to decide on the desirable statistical form for the compilation of the actual data for decision-

1 Is the information required:
   (i) to decide on location;
   (ii) to assess a shape;
   (iii) to decide on orientation;
   (iv) to assess the suitability of a design under statistically defined extreme conditions;
   (v) to assess a running condition;
   (vi) to plan operation on a temporal basis?
2 Precisely what climatic elements enter into the problem? Is the optimal condition known?
3 Are simultaneous values of several climatic variables required? If so, are they likely to be available?
4 Is the temporal variation of relevant factors important? Are diurnal or seasonal variations of values of most significance, or both?
5 Is the spatial variation of relevant factors important? If so:
   (i) are horizontal variations important? At what scale?
   (ii) are vertical variations important? At what scale?
6 Does interest centre on:
   (i) a statistically defined extreme condition? Is this dependent on a single element?
   (ii) an average condition? Is this dependent on a single element?
7 If interest centres on a statistically defined extreme condition what will be the consequences if the selected design values are exceeded?
   What factor of safety is needed?
   Alternatively, with what frequency can the possibility of failure be tolerated without serious consequences?
8 If interest centres on an average condition, over what period of the year should the average be taken, eg, what is the length of the heating season?
9 What information is actually available, and will this supply an answer that can actually be used for design purposes?
10 How representative of the place actually under consideration are the records of the station to be used?
11 What additional information might be made available
   (i) from a study of existing records?
   (ii) by adding new observations?
12 Can new observations produce results within the time period available?
   Can short-term records be linked to long-term records, if special information is required?
13 If new information is requested, precisely how would it be used for design purposes?
14 Are there any microclimatological site considerations that need special attention or special study?

Fig. 16.1 A check list for using climatological data for urban design purposes

making. For example, in design for structural stability against wind failures, a mean recurrence period of fifty years is in current use in the UK. The problem is therefore to forecast the once-in-fifty-year gust velocity. The required meteorological data may be linked with extreme values. Alternatively, average values may be more appropriate or perhaps frequency distributions of associated values of several variables over the whole range of data, both on a daily and a seasonal basis. Many of the practical problems are multi-element problems, therefore statistical combinations of several variables are frequently necessary, for example in all heat transfer problems (Page, 1970). However, we have reached the stage in analysis when it should be possible, given an adequate meteorological network and computer data bank, for a meteorological service to compile the relevant climatological data for the green-field site on an adequate statistical base. Unfortunately, these data are not what we want for urban design purposes, for we cannot have 'green-field' towns.

Clearly the macrometeorological data has to be modified to allow for urban effects before it can be fed into any actual design problem.

Three types of urban modification can be identified:

(a) Local climatic modifications due to terrain, slope climates, etc.

(b) Generalised, wide-scale, urban climatological modifications representing the statistical effects of the urban complex integrated over an appreciable section of the urban space, eg, heat island effects, etc.

(c) Local urban microclimatological modifications, resulting from detailed urban form on and around a particular site, like over-shadowing, wind channelling, etc.

It is easier to state the problem than to solve it. Landsberg (1970) has established the broad features of urban climates, but such data are fairly vague to apply to precise design situations (Table 16.1). We clearly need to be able to construct theoretical models that enable us to predict the influence of all the above factors simultaneously, but, in view of the complexities of urban form, this is well-nigh impossible. Therefore some simplification becomes necessary. One approach is to assume that the terrain correction can be estimated separately from the other two corrections. The validity of this approach will depend on the relative scale of the urban structure and the

Table 16.1

Element	Comparison with Rural Environs
Contaminants:	
dust particles	10 times more
sulphur dioxide	5 times more
carbon dioxide	10 times more
carbon monoxide	25 times more
Radiation:	
total on horizontal surface	15 to 20% less
ultraviolet, winter	30% less
ultraviolet, summer	5% less
sunshine duration	5 to 15% less
Cloudiness:	
clouds	5 to 10% more
fog, winter	100% more
fog, summer	30% more
Precipitation:	
totals	5 to 10% more
days with less than 5 mm	10% more
snowfall	5% less
Temperature:	
annual mean	0 to 1·0°F more
winter minima (average)	1 to 2°F more
heating degree days	10% less
Relative humidity:	
annual mean	6% less
winter	2% less
summer	8% less
Wind Speed:	
annual mean	20 to 30% less
extreme gusts	10 to 20% less
calms	5 to 20% less

Sources: H. E. Landsberg: Climates and urban planning in WMO Technical Note No 108, 1970.

also H. E. Landsberg: City air—better or worse, in *Symposium on air over cities*, R. A. Taft, Sanitary Engineering Centre, Public Health Service, SEC Report No A62–5, 1961.

terrain. The urban influence correction can then be applied, and finally the micrometeorological modification. The urban influence, however, is not uniform and is a function of urban position as well as urban scale. It is therefore necessary to know not only the magnitude of the maximum urban green-field difference, but ideally also the spatial description of the pattern, in mathematical terms. For example, a heat island is related to urban form, scale, and density by some relatively continuous temperature variation from the rural area to the centre. Is the gradient linear for a circular city, or is it a more complex function, say exponential?

Forecasting the Climatological Properties of the Built Environment
The most important practical urban meteorological modifications relevant to design decision-making would appear to be as follows.

(a) Modification of the wind-flow field due to the drag and other boundary layer disturbances exerted by the urban form.

(b) Modification of the distribution of short-wave radiation pattern due to the geometry of the town, and the modification of the transmission characteristics of the atmosphere by the pollutants put out by that town.

(c) Modification of the net radiation pattern due to a number of factors, including urban geometry, atmospheric pollution, thermal storage in structures, energy inputs from heating appliances, transportation, etc. A preliminary discussion of the net radiation problem has been given by Fuggle and Oke (1970).

In order to develop a forecasting technique, it is necessary to develop relationship functions to find the modified climate from a description of urban form as measured by some appropriate urban design parameters, for example, mean roughness height of the development is relevant to the study of the build up of the urban wind boundary layer. Proceeding in this way, it is possible, for example, to estimate the boundary layer profile as the wind moves into an urban area (Munn, 1970). The wind problem has received fairly extensive study recently (Royal Society, 1971), but we are still a long way from internationally agreed procedures for forecasting urban wind fields. Most knowledge probably exists on the short-wave radiation problem, but it is the net radiation pattern which is so critical to the actual thermal outcome in towns. Here we are in an area of considerable ignorance. The magnitude of the various terms in the net radiation balance has not yet been established with any certainty, and we are even farther from knowing how to predict the net radiation from descriptions of urban geometry, thermal heating outputs, typical thermal characteristics of urban materials, etc. Forecasting the thermal climates of towns is thus a very pragmatic art, and will remain so if we cannot reduce the problem to a better physical basis linked with energy considerations.

Micrometeorological Modifications at Site Scale
Finally, having found the unmodified climate of the urban site, it is necessary to decide how that climate is likely to be modified at the micrometeorological level by the actual construction erected on that site and the other construction systems erected around that site. In some cases this problem can be tackled by physical model building techniques, eg, in aerodynamic studies by using wind tunnels. Other aspects of the problem cannot be easily handled in this way, for example, the thermal climates of courtyards in a wind field. Here, at present we have to revert to the classical micrometeorological technique of analogous situations described systematically in the literature. This is not a very satisfactory technique, but this is all we have for the majority of urban site forecasting situations. Model techniques are useful, but are relatively inflexible. Hence they frequently have less value than might first appear in the relatively fluid situation in which day-to-day design takes place, which involves the rapid two-dimensional consideration of a large number of possible three-dimensional alternatives. Theoretical relationship functions are much to be preferred in design, which is essentially a time-limited theoretical activity aiming to produce the information needed for reaching a desired practical outcome as a process of optimisation between a large number of variables bearing on the design, only one of which is climate.

The WMO/WHO Brussels Conference on Urban Climatology, and Building Climatology, which was also supported by CIB provided a good summary of current state of knowledge in these fields (WMO, 1970, a and b). Further information may be found in standard texts dealing with urban climatology, like Geiger (1959) and Chandler (1965). Interesting methodological approaches are considered in the Russian literature, for example in Gol'tsberg's book (1967). There is, however, a serious lack of knowledge of vertical variations in the urban climate, for example of temperature gradients.

A Preliminary Mathematical Approach to Forecasting the Urban Climate

Basically one is first seeking relationship functions of the form:

$$W_1 = f(w_1, P_1, P_2, P_3, \ldots) \tag{2}$$

where W_1 is the value of the site weather variable in the absence of the project

w_1 is the value of the corresponding macrometeorological weather variable from the nearest meteorological site

$P_1, P_2, P_3, \ldots$ are the properties of the city, topography and landscape producing modification of the urban environment

Sometimes these relationship functions emerge in the form:

$$W_1 = w_1 + x_p \tag{3}$$

where x is an arithmetical correction expressed as a function of some physical variable characteristic of the town, say, scale or density for a position p in the urban complex.

It would appear to be worth pondering whether such urban climate modification functions of the form:

$$W_1 = w_1 + x$$

can have universal validity. For example, Ludwig (1970) studied urban temperature fields with special reference to urban–rural temperature differences. He found a relationship between lapse rate $\gamma°C/mb$, and Δt, the urban–rural temperature difference, $°C$, of the form:

$$\Delta t = 1·85 - 7·4\gamma \qquad \text{correlation } -0·79$$

He also studied the effect of urban size, dividing the data into three groups, (1) smallest five, (2) next five, and (3) two largest cities (London and St Louis).

He obtained the following regressions:

Group (1) $\Delta t = 1\cdot3 - 6\cdot78\gamma$ correlation $-0\cdot95$

Group (2) $\Delta t = 1\cdot7 - 7\cdot24\gamma$ correlation $-0\cdot81$

Group (3) $\Delta t = 2\cdot6 - 14\cdot8\gamma$ correlation $-0\cdot87$

The successful combination of the American and English data may depend on similarities of urban pattern, and energy balance, but such similarities might not be universal. It would appear that considerable caution is needed in carrying regression functions from one area to another in the present pragmatic state of the art.

For other weather factors, for example wind, the relationship function may take the proportional form:

$$W_1 = kw_1 \tag{4}$$

where k is a constant derived from appropriate physical characteristics of the town and the site.

For example, in wind loading in the UK, the basic design wind velocity is modified by two multiplying factors, S_1 and S_2, to allow for the effects of site terrain and structure height (British Standards Institution, 1970).

Responsibility for action in this area of study should rest primarily with the meteorologist, but he must see his relationship to the total design decision chain, because the outputs of meteorological study must be in a suitable form to become the inputs into the more detailed, building-climatological study of a particular site in its environmental context.

A Preliminary Mathematical Approach to the Forecasting of the Detailed Microclimate of the Site

An important research problem in outdoor building climatology is the derivation of relationship functions that enable the properties of the building as a climatic modifier to be evaluated, so that the relevant microclimate can be predicted from the macroclimatic data, as modified by the urban influence.

Mathematically one could consider the relationship function to derive the local environment from a set of simultaneously forecast weather variables for unobstructed sites, $W_1, W_2, W_3, \ldots$, etc (expressed as vectors on the macroclimatic scale), interacting with a building with a set of geometrical properties $P'_1, P'_2, P'_3, \ldots$, etc, as being of the form:

$$(E_1, E_2, E_3, \ldots)_L = f(W_1, W_2, W_3, \ldots, P'_1, P'_2, P'_3, \ldots) \tag{5}$$

where $E_1, E_2, E_3, \ldots$ are the resultant associated local microclimatological variables forecast at point L. Sometimes the climatological variables can be separated, and sometimes they cannot.

Quite often E_1 is found to be a linear function of W_1, and the relationship function can be reduced to the form:

$$(E_1)_{L\theta} = W_1 f_\theta(P'_1, P'_2, P'_3, \ldots) \tag{6}$$

where θ indicates that the relationship holds for the weather variable from a direction θ.

Wise (1970), for example, has studied the relationship function for maximum environmental wind velocity V_A at height A between interacting buildings for the wind perpendicular to the buildings, and expressed it in the following proportional form: to calculate V_A from V_H, the mean wind velocity in free stream at height H:

$$\frac{V_A}{V_H} = 0.24 \left\{ \left(\frac{a}{H}\right)^{0.28} + \left(\frac{L}{H}\right)^{0.4} \left(\frac{W}{H}\right)^{0.4} \left(\frac{H}{h}\right)^{0.8} \right\}$$

where the terms on the right-hand side describe the geometry of the interacting buildings. It is implicit in this expression that the relationship function is independent of Reynolds number.

Some relationship functions are very simple to solve in terms of building geometry, for example, absence or presence of shade at a given time of year at a given place, or the direct energy reaching an exposed surface from the sun. Others are much more complicated to evaluate. For example, it is known that tall buildings bring down large volumes of air from above surface to pedestrian levels. What are the simultaneous modifications of the ground-level air temperature and wind velocity which take place? These data might be very relevant, for example in assessing outdoor cold stress, using the wind chill index.

A major task in urban forecasting is going to be the building-up of such relationship functions into a comprehensive design set. The problems will be similar to those encountered in deriving equivalent relationship functions for wind loading, but more complex in view of the greater number of physical variables involved.

Microclimatic Design—Practical Considerations

The data generated by the above systematic approach would be very extensive, and some simplification would be necessary for practical design. This would involve identifying the location of the site positions and urban locations where the design variable has a critical importance. For example, in order to avoid having to explore the whole spatial-design wind-field in detail, a simplification is often introduced. E_{max} is treated as the design variable, eg, one might state that the wind speed at the worst point should not exceed a certain value for more than a certain proportion of the year or season. Unfortunately, the position of E_{max} is liable to vary with the direction of the weather variable. The study of the protection diagrams provided by isolated building with the wind from different directions made by Jensen and Franck (1963) shows that the apparently simple relationship function described above may be very complex in detail.

A possible approach to design is to attempt to identify known achievable microclimates of good general characteristics associated with established building forms known to have favourable properties, and then to compare predicted microclimates for other building forms with the established good achievable microclimates. Such an approach recognises the role of optimisation in environmental design, and helps avoid the setting of unachievable goals. For example, Wise, Sexton, and Lillywhite (1965) of the UK Building Research Station have demonstrated that the environmental wind at ground level in between traditional rows of building is likely to be of the order of half the freewind speed at 10 m. With tall buildings interacting with low buildings, the environmental wind may reach double the free wind (Wise, 1970). This represents a wind deterioration ratio of four due to adverse design, or a wind force deterioration ratio of 4^2, ie, 16. It is no wonder that the public have a poor opinion of the contemporary architect's achievements in the control of outdoor wind.

The permissible increases in the target microclimates could be made dependent on the basic characteristics of the site climate. For example, on a very windy site, the permissible wind increase ratio above target level should be kept very much lower than in a very sheltered area.

Conclusions

The author is aware both of the lack of available techniques for urban climatological forecasting, and also of the need for a fuller and comprehensive review of this difficult field. It is nevertheless hoped that these ideas will stimulate interest in systematic urban climatology and demonstrate the importance of improving urban climatological forecasting. It is always likely to remain a difficult art, but priority must be allocated to further study of the net radiation balance of towns to clarify the relative magnitude of the various terms in the heat balance equation. Pollution has not been taken as a meteorological variable, because the pollution climate is the consequence of the various pollution outputs in relation to the urban climate, which has to be forecast first in any model studies of pollution. Vertical variations of climate are particularly important in the pollution problem, but unfortunately there is remarkably little information that would enable one to forecast the vertical climate of towns. There is little enough vertical information anyway for the green-field site itself.

References

BRITISH STANDARDS INSTITUTION (1970). *Code of basic data for the design of buildings*, Chapter V, Loading, Part 2, Wind loads. *B.S.I.*, London, 12–13.

CHANDLER, T. J. (1965). *The climate of London*. Hutchinson, London.

FUGGLE, R. F. and OKE, T. R. (1970). Infra-red flux divergence and the urban heat island, in *Urban climates*. Technical Note No 108, WMO, No 254, TP 141, WMO, Geneva, 70–78.

GEIGER, R. (1959). *The climate near the ground*. Harvard University Press, Cambridge, Mass.

GOL'TSBERG, L. A. (1967). *Microclimate of the U.S.S.R.* Gidrometeorologicheskoe Izdatel'stvo, Leningrad, 1967, translated English, Israel Program for Scientific Translations, Jerusalem, 1969.

JENSEN, M and FRANCK, M. (1963). *Model scale tests in turbulent wind*, Pt. I (English). Danish Technical Press, Copenhagen.

LANDSBERG, H. E. (1970). Climates and urban planning, in *Urban climates*. Technical Note No 108, WMO, No 254, TP 141, WMO, Geneva, 364–71.

LUDWIG, F. L. (1970). Urban temperature fields, in *Urban climates*. Technical Note No 108, WMO, No 254, TP 141, WMO, Geneva, 80–107.

MUNN, R. E. (1970). Air flow in urban areas, in *Urban climates*. Technical Note No 108, WMO, No 254, TP 141, WMO, Geneva, 15–39.

PAGE, J. K. (1970). The fundamental problems of building climatology considered from the point of view of decision making by the architect and urban designers, in *Building climatology*. Technical Note No 109, WMO, No 225, TP 142, WMO, Geneva, 9–21.

ROYAL SOCIETY (1971). *Phil. Trans. Roy. Soc.*, Series A, **269,** Part 1199, 321–554.

THOM, H. C. S. (1970). Application of climatological analysis to engineering design data, in *Building climatology*. Technical Note No 109, WMO, No 255, TP 142, WMO, Geneva, 233.

WISE, A. F. E. (1970). *Wind effects due to groups of buildings*. Building Research Station, Garston, England, Current Paper No. 23/70.

WISE, A. F. E., SEXTON, D. E., and LILLYWHITE, M. S. T. (1965). *Air flow round buildings*. Urban Planning Research Symposium, Building Research Station, Garston, England.

WMO (1970a). *Urban climates*. Technical Note No 108, WMO, No 254, TP. 141, WMO, Geneva.

WMO (1970b). *Building climatology*. Technical Note No 109, WMO, No 255, TP 142, WMO, Geneva.

CHAPTER SEVENTEEN C. V. BARNETT

Weather and the Short-Term Forecasting of Electricity Demand

Introduction

At the flick of a switch or the press of a button the electricity consumer has instantaneous power at his command. This continuing miracle is apt to be taken for granted, for it is not always realised that electricity cannot be stored in appreciable quantities but can be generated only for *immediate* use. A few degrees fall in temperature or a sudden thunderstorm darkening the sky over a densely populated area, and the increased demand felt throughout the country may exceed 1,000 MW. Such changes in weather conditions must be anticipated, if possible, by the electricity supplier, for it can take three hours or more to put an additional large generator on load. Thus accurate weather forecasts are vital to the smooth operation of the National Grid System whilst maintaining only a minimum of spinning reserve capacity.

Without the variability of the weather to contend with, it would be a simple matter to forecast future electrical demand in the short-term. The CEGB system demand would then show a marked regular pattern of daily, weekly, and seasonal variation. However, this regular pattern is to a large extent masked by the erratic behaviour of the weather which accounts for practically the whole of the variability in demand (Davies, 1958).

History

A method of forecasting demands using weather adjusting techniques was first reported in 1944, in a paper read before the American IEE, relating to a successful application by the Philadelphia Electric Company. Later in that year a similar method was instituted in England, in the predominantly non-industrial St Paul's Grid Control Area (now St Albans and Grinstead Grid

Control Areas), where accurate demand forecasting has proved impossible by ordinary projection methods. During 1949 the use of the method was extended to the other Control Areas. Since then the story has been one of gradual improvement in methods; introducing empirical statistical analysis of demand and weather data, using specially derived weather functions, and gradual computerisation.

Operational Requirements at National Control

The CEGB operates a three-tier system of control of the grid network in England and Wales. There are seven Grid Control Centres (see Fig 17.1) responsible for the Supergrid system in their Area (275 and 400 kV) and each Grid Control Centre has associated with it one or more District Centres which are responsible for the local network at lower voltages (132 kV and lower). Co-ordination of the Grid Control Centres and planning of national resources is carried out by the National Control Centre in Central London.

The time scale for which load forecasts are made by Grid Control Areas and National Control ranges from up to two years for generation and transmission outage programmes down to three hours ahead for Control Room purposes.

National Control is responsible for forecasting for the whole of England and Wales:

(a)   Winter peak demands under average-cold-spell (ACS) weather conditions for the next two years (see p 221).

(b)   Weekly peak demands under ACS weather conditions for up to three years ahead for preliminary outage programmes.

(c)   Weekday peak demands under expected weather conditions up to three weeks ahead for dynamic planning of the generation outage programmes.

(d)   Weekend peak demands for up to two weeks ahead.

(e)   Weekend and Bank Holiday minimum demands from weekend ahead to up to five months ahead.

(f)   Summer minimum demands for the next two years.

(g)   The general, daily load-shape and any emerging changes that may require amendments to costing periods. (The twenty-four hours of the day are divided into specific periods during which the marginal cost of producing electricity is assumed to be constant.)

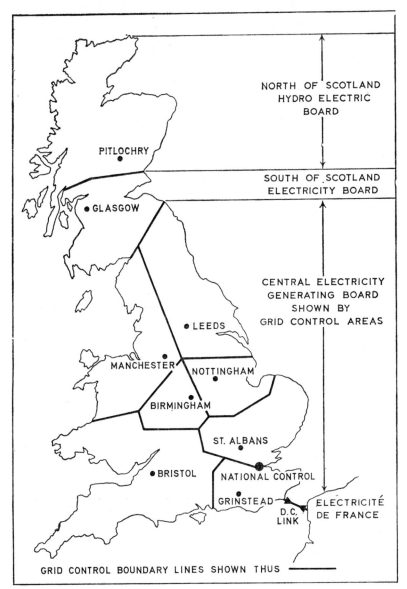

Fig. 17.1 Map of the CEGB Grid Control Areas

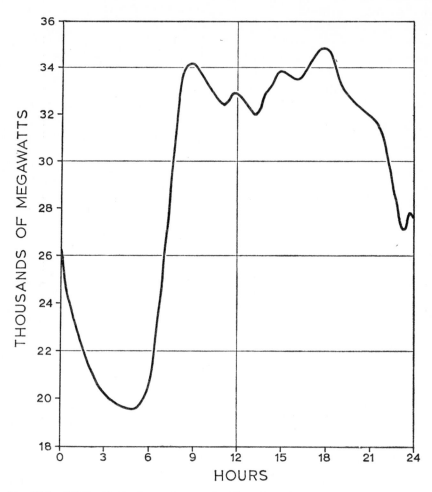

Fig. 17.2   CEGB: Daily demand curve for Thursday 14 January 1971

The Daily Demand Curve

An example of a daily demand curve is shown in Fig 17.2. The day in question, 14 January 1971, was a typical winter weekday. Demand is low during the early part of the morning, and consists mostly of heavy industrial loads (twenty-four hours continuous process industries) and domestic storage heating. As the morning proceeds, activity in domestic and light industrial fields builds up to a morning peak between 08.00 and 10.00 hours. Domestic activity falls away after breakfast but this is counteracted by the increasing commercial demand so that total demand falls only slightly until the lunchtime increase between 12.00 and 12.30 hours. Industrial load falls away slowly during the afternoon but the domestic storage-heating boost creates another peak in demand between 14.00 and 15.00 hours. A further maximum occurs at about 18.00 hours and this is due to the domestic load building up again as people arriving home switch on the lights and cook the evening meal. Demand now gradually falls away again as evening activity subsides until the off-peak heating returns at 23.00 hours and then from midnight onwards falls away rapidly till 03.00 hours.

At other seasons the pattern of demand varies slightly, eg, during the summer the darkness peak occurs as late as 22.30 hours and is quite distinct from the tea-time peak at 17.30 hours. The change-over from BST to GMT and vice-versa also affects the demand curve.

Meteorological Data for Day-to-Day Operation

Demand forecasts are submitted by the Grid Control Centres for each of the costing periods of the daily demand curve. The forecasts are based on special regional short-range weather forecasts supplied by the Regional Meteorological Offices to the respective Grid Control Centre. Both provisional and final weather forecasts are provided, the former being issued seven to fourteen hours, and the latter three to five hours, in advance of the event. The sum of these individual Grid Control Area estimates provides one means of obtaining a global CEGB demand estimate.

Forecasts of the total demand on the system are made each afternoon by National Control to cover the main costing periods of the following day. These forecasts are updated on the day in question in the light of further actual demand and weather experience and revised weather forecasts. These demand forecasts are based on weather forecasts supplied by the Regional Meteorological Offices through the respective Grid Control Centres and collated at National Control. These updated National Control estimates are compared with those referred to in the preceding paragraph.

Details of the weather forecasts submitted by the Regional Meteorological Offices are as follows:

1 *Day Ahead Forecast* issued at 14.00 hours covering forecasts 06.00–09.00, 09.00–12.00, 12.00–15.00, 15.00–18.00, 18.00–21.00 hours.

2 *Preliminary Forecast* issued at 18.30 hours covering forecasts 06.00–09.00, 09.00–12.00, 12.00–15.00 hours, plus lightning risk 18.30–06.00 hours and outlook 15.00–24.00 hours.

3 *Final and Revised Forecasts* issued at 04.30 hours covering forecasts 06.00–09.00, 09.00–12.00 hours, plus lightning risk 04.30–12.00 hours and outlook 12.00–24.00 hours.

4 *Final, Revised, and Preliminary Forecasts* issued at 08.30 hours covering forecasts 09.00–12.00, 12.00–15.00, 15.00–18.00 hours, plus lightning risk 08.30–18.00 hours and outlook 18.00–06.00 hours.

5 *Final and Preliminary Forecasts* issued at 11.00 hours covering forecasts 12.00–15.00, 18.00–21.00 hours, plus lightning risk 11.00–21.00 hours and outlook 21.00–09.00 hours.

6 *Final Forecast* issued at 13.30 hours covering forecast 15.00–18.00 hours, plus lightning risk 13.30–24.00 hours and outlook 21.00–09.00 hours.

7 *Final Forecast and Actuals* issued at 16.00 hours covering forecast 18.00–21.00 hours plus lightning risk 16.00–24.00 hours and outlook 24.00–15.00 hours.

8 *Long-Term Weekend Forecast* issued by 16.00 hours on Thursday covering maximum and minimum temperatures plus weather indications for Saturday, Sunday, and Monday.

The weather forecasts include information on the following parameters:

1 Temperature: the average for the three hours covered by the forecast.

2 Wind: direction and speed (knots).

3   Cloud: height and amount of main layer using the scale:

below 2000 feet (610 m)

2000–5000 feet (610 m–1524 m)

above 5000 feet (1524 m)

total cloud amount (in eighths) and approximate vertical thickness (thin, medium, or thick).

4   Precipitation: type, intensity, frequency, and distribution.

5   Visibility: in yards or miles with fog indicated.

Hourly weather reports are made at a large number of observing stations in this country by the Meteorological Office. The London Weather Centre provides National Control each day with synoptic reports from twenty weather stations covering hourly temperatures and cloud and other information at 09.00, 12.00, and 17.00 hours. Reports of cloud and other information are also given for 06.00, 15.00, and 21.00 hours for seven weather stations.

In addition to the above reports National Control also receive hourly plotted charts of the weather fifty minutes after the event at over a hundred meteorological stations by means of the Landline Facsimile Ring Circuit, known as the Weatherfax Service, emanating from the Meteorological Office Headquarters at Bracknell, Berkshire. These are especially useful in deducing the likely arrival of adverse weather conditions at large conurbations, and in extrapolating weather trends for updating demand forecasts.

For each specific meteorological factor investigated it is necessary to calculate a weighted average of the simultaneous values at reporting stations selected according to the geographical distribution of load density. The weighting is proportional to the weather-sensitive component of the demand in the region of which each reporting station is considered representative, the weights being in the first instance established by analyses of each Grid Control Area separately.

Originally, twenty meteorological stations were used to calculate these weights but it was found that weather relative to the whole system could be represented to sufficient accuracy by only six stations.

Effect of Weather on Demand

The meteorological elements found to affect the demand are as follows (Davies, 1958):

(a)   Temperature

(b)   Wind speed

(c)   Cloud

(d)   Visibility

(e)   Precipitation

Elements (a) and (b) control the heating demand, the wind speed being allied to temperature in dissipating heat from buildings. (c), (d), and (e) are used to estimate the level of daylight illumination which determines the lighting demand. Precipitation can also reduce the diversity by keeping people indoors and thus increase demand.

The meteorological information listed above is processed into the following 'working' variables:

(i)   'Effective' temperature (x_1)

(ii)   Cooling power of the wind (x_2)

(iii)   Daylight illumination index (x_3)

(iv)   Rate of precipitation (x_4)

Effective temperature takes account of the time-lag of consumer response to temperature variations due to thermal storage in the fabric of buildings. The simplest equation which expresses the lag of internal temperature, θ, is the first-order lag equation:

$$(d\theta/dt) + \lambda\theta = \lambda T_0$$

where $1/\lambda$ is the thermal time constant and T_0 is the ambient temperature. If t is measured towards the past and θ is the room temperature at time $t = 0$, the solution of this equation (ignoring the transient term) is:

$$\theta = \lambda\int_0^\infty e^{-\lambda t} T_0(t)\, dt.$$

The next approximation is to replace the integral by summation over hourly intervals:

$$\theta = \lambda(T_0 + e^{-\lambda}T_{-1} + e^{-2\lambda}T_{-2} + \ldots)$$

where $T_0, T_{-1}, T_{-2}, \ldots$ are the external air temperatures at time $t = 0, 1, 2, \ldots$ hours. This equation can be rewritten

$$\theta = \alpha(T_0 + \beta T_{-1} + \beta^2 T_{-2} + \ldots)$$

where $\beta = e^{-\lambda}$, $\alpha + \beta = 1$, and account has been taken of the need for $\theta \to T_0$ as all temperatures become equal.

Calling θ the effective temperature and writing T_E, T_{E-1}, $T_{E-2}, \ldots$ for its value at time $t = 0, 1, 2, \ldots$ the above equation is equivalent to:

$$T_E = \alpha T_0 + \beta T_{E-1}$$

This equation has the disadvantage of requiring all twenty-four hourly temperatures each day, so the final simplification is to make T_0 the average temperature in the three hours preceding the event, the assumption being that the temperatures immediately before the event in question and at similar times on previous days are of greatest relevance to the level of demand. With this assumption a value of α of $\frac{1}{2}$ is found to correlate closest with demand. Thus, effective temperature is finally defined by the equation:

$$T_E = (T_0 + T_{E-24})/2.$$

Cooling power of the wind gives the additional thermal effect due to wind in association with temperature. As a result of a number of statistical investigations (Davies, 1958), the best formula for the cooling power of the wind has been found to be

$$W^{\frac{1}{2}}(18 \cdot 3°C - T)$$

where W is the wind speed in knots and T the ambient temperature in °C.

The daylight illumination index is defined by

$$100 \log_{10} I$$

where I is the daylight illumination in kilolux on a horizontal surface exposed to sun and sky as estimated from reports of cloud, visibility, and precipitation. This calculation is based on an analysis of seven years of observations of daylight illumination at Kew. The analysis furnished basic data regarding the illumination on perfectly clear days as a function of the sun's elevation, data on the transmission factor of individual cloud layers, and data on the transmission of daylight through the atmosphere near the ground as a function of visibility. From this basic data the transmission through the cloud layers present can be calculated. While the relationship between lighting demand and daylight illumination is only approximately logarithmic, it has been found useful to retain the unit.

Precipitation is a broad classification for rain, drizzle, snow, hail, sleet, etc, and its effect on the demand is partly to reduce the daylight illumination. Although in theory, it should also increase the rate of heat loss from the surfaces of buildings, in fact, the only other effect of precipitation seems to be that of keeping people indoors. This is suggested by the fact that only the evening peak appears to be directly affected by rainfall. Average rainfall in millimetres per hour at the time of peak is included in the analysis only in sufficiently wet periods.

Demand Analysis and Forecasting

Demand Analysis

Weekdays, weekends, and holiday periods are analysed and forecast separately for each significant peak and trough demand in the daily load curve. A model of the following form is used for analysis and forecasting purposes:

Demand = Basic Level + Weather Demand

+ Day of Week Correction + Random Component

The coefficients in the model are obtained from a regression analysis of the corresponding period of the previous year and these are adjusted for growth.

The method of demand analysis is based on the assumption that the weather-sensitive component of demand can be expressed as a sum of functions of the respective meteorological factors as shown below:

$$Y_r = b_0 + b_1 x_1 + b_2 x_2 + b_3 x_3 + F(r) + b_5 D + E_r$$

where Y_r is the demand at a fixed period on day r; x_1, x_2, x_3 are the corresponding specific meteorological factors; $F(r)$ is a polynomial function of the time of year; D is a day of week variable, b_0, b_1, b_2, b_3, b_5 are constants, and E_r is the unexplained or residual component of demand. In this expression the known quantities are the dependent variable Y and the independent variables x_1, x_2, x_3, r, and D. The coefficients in the model above are found from a regression analysis of the data. Further meteorological variables may be added to this expression, or existing ones deleted, as required. Thus, in the analysis of the evening peak demands the variable x_4 may be incorporated while, in the analysis of demands in darkness, x_3 is not used.

The assumption of a linear relationship implies that the regression analysis will yield only average effects over the range covered by the data. It would be possible to allow for a non-linear relationship by introducing more complex mathematical functions. Thus it is found to be useful in some analyses to include a quadratic temperature function.

The sensitivity of demand to weather changes can be illustrated by the following figures for the winter period:

(a) Temperature—a fall in temperature of 1°C will increase the demand by about 1·8 per cent.

(b) Wind—a change in wind speed from 4 knots to 9 knots will increase the demand by about 1 per cent assuming temperature does not change.

(c) Cloud—a change in lighting conditions from a clear to a half-obscured sky on a dry day would cause an increase in demand of about 1·4 per cent, assuming there are two cloud layers present. On a wet day the response would be doubled.

It should be noted that the above sensitivities are average levels for a typical winter weekday. Sensitivity to weather has a diurnal as well as seasonal pattern.

Demand Forecasting

In applying the results of the analysis of the corresponding period of the previous year to day-ahead forecasting, an estimate is made of the growth of the basic level using recent historic demand and weather data. Forecast weather, suitably processed, is then weighted by the relevant MW responses and a day-of-week correction applied.

The simplest model for which satisfactory demand forecasting can be obtained has the following equation:

Actual Load $=$ Basic Load $+$ Weather Correction

$+$ Day-of-Week Correction

or
$$L = a_0 + a_1 r + a_2 r^2 + \ldots + b_1 T + b_2 I + b_3 C + \sum_j \delta_{ij} D_j$$

where

$L =$ actual load

$r =$ day number, eg, $r = 1$ for first day, $r = 2$ for second day, etc

$T =$ effective temperature

$I =$ effective light

$C =$ cooling power of the wind

$i =$ day-of-week number, ie, $i = 1$ on Monday, $i = 2$ on Tuesday, etc

$\delta_{ij} = 1, i = j$

$\quad = 0, i \neq j$ (j is summed from 1 to n, where n is the number of days of the week involved in the analysis, eg, if weekdays only are involved, then $n = 5$)

$D_i =$ day-of-week correction for day i

The above model has the disadvantage that it uses no demand information within about twenty-four hours of the time being forecast and since there is considerable correlation between the levels of successive peaks and/or troughs in the demand profile, improved estimating is often obtained by using the following model:

$$L = b_0 L_P + a_0 + a_1 r + a_2 r^2 + \ldots$$

$$+ b_1 T + b_2 I + b_3 C + c_1 T_P + c_2 I_P + c_3 C_P + \sum_j \delta_{ij} D_j$$

where

$L_P =$ load at previous peak (say)

$T_P =$ effective temperature at previous peak, etc

A further improvement in demand forecasting is often possible by considering the ratio of successive peaks (the so called multiplier system) in the form:

Actual Multiplier = Basic Multiplier

+ Weather Correction + Day-of-Week Correction

or $L/L_P = a_0 + a_1 r + a_2 r^2 + \dots$

$$+ b_1 T + b_2 I + b_3 C + c_1 T_P + c_2 I_P + c_3 C_P + \sum \delta_{ij} D_j$$

This method also has the advantage that usually some of the coefficients are found not to be significant and can be removed from the equation.

The high correlation which often exists between the various so-called 'independent' variables used in the analysis of demand/weather relationships can lead to unsatisfactory least-squares regression coefficient estimates. More satisfactory results are obtained by Ridge Regression techniques (Hoerl and Kennard, 1970) which are finding application in this work.

Standard Weather Demands

For planning and operational purposes, historic and forecast demands are expressed in terms of the following standards:

(a)   normal weather demands

(b)   average-cold-spell (ACS) weather demands

(c)   extreme severity weather demands

Demands in normal weather are readily understood, but further explanations are required in the interpretation of average-cold-spell and extreme severity demands and weather. ACS weather has its roots in the concepts of ACS demands, and is arrived at indirectly by considering weather associated with weekly or yearly peak demands. It must be emphasised that the standards of ACS weather currently used by National Control are not unique, but only one of an infinite number of combinations of relevant weather variables which would satisfy, in terms of peak demands, the appropriate risk criterion. It is the aggregate MW effect of all significant weather variables on the underlying level of demand for that time-of-day, day-of-week, and time-of-year, which is important.

J

The only official definition of ACS demand refers to the winter as a whole. In paragraph three of Appendix 70 of the Report of the Select Committee on Nationalised Industries (28 May 1963) it is stated that 'the planning of generating capacity is related to the peak load to be expected on a winter weekday in average-cold-spell weather. This is determined by calculating the yearly peaks that would correspond to the range of weather likely to be experienced in a span of 100 years and selecting the median. Thus, in 50 years the peak demand would be higher, and in 50 years lower, than the demand in average-cold-spell weather.'

Although extreme-severity weather-demand is not officially defined the probability of experiencing such a demand has been taken into account in assessing the planned, gross plant margin of 20 per cent. The extreme-severity weather-demand is that winter peak demand which is exceeded in only 1 per cent of the occasions. Its level has been put at 9 per cent above the winter ACS demand as the result of the experience of the severe winter of 1962/63.

There is no official definition of weekly ACS demand although a set of weekly ACS weather has been used by National Control for some years. These revised standards of weekly ACS weather have a uniform weekly risk of weather being worse and are also linked with risks on an all-winter basis on which plant programmes are based. It would appear that what would be a 50 per cent risk on an all-winter basis is about a 12 per cent risk in the peak week of the year. This is not unexpected since there is a greater risk of severe weather occurring sometime during the winter than in a particular week.

Conclusion

In the case of three-hours-ahead demand-forecasting, the methods described in the foregoing paragraphs normally result in estimates within 1·5 to 2·0 per cent of actual demand (standard deviation). The total error can be approximately equally divided between model error and weather forecasting error. It is unlikely that any significant improvements can be obtained in the model used without introducing further complexities. It is hoped, however, that the quality of the weather forecasting service will improve as new, computer-aided techniques become available.

References

DAVIES, M. (1958). *The Relationship between Weather and Electricity Demand.* I.E.E. Monograph No. 3145.

HOERL, A. E., and KENNARD, R. W. (1970). Ridge Regression: Biased Estimation for Nonorthogonal Problems. *Technometrics*, **12**, No. 1, 55.

Selected Classified Bibliography

References given at the end of chapters have, for the most part, been omitted from this bibliography to avoid repetition. It has been classified for convenience of reference but it should be emphasised that many items overlap into several categories. (The editor gratefully acknowledges the co-operation of his colleague, Mr D. J. Unwin, in the compilation of this bibliography.)

1 Agricultural/Biological

BEAUCHAMP, E. G. and LATHWELL, D. J. (1967). Root-zone temperature effects on the early development of maize. *Plant and Soil*, **26**, No 2, 224–34.

BULL, T. A. (1968). Expansion of leaf area per plant in field bean (*Vicia falia* L) as related to daily maximum temperature. *J. Appl. Ecol.*, **5**, 61–8.

CURRY, L. (1962). The climatic resources of intensive grassland farming; The Waikato, New Zealand. *Geog. Rev.*, **52**, 174–94.

CURRY, L. (1963). Regional variation in seasonal programming of livestock farms in New Zealand. *Econ. Geog.* **39**, No 2, 95–118.

DENT, J. B. and ANDERSON, J. R. (1971). *Systems analysis in agricultural management*. Chapters 7, 8, 12, and 14. Wiley, London.

DUCKHAM, A. N. (1963). *Agricultural synthesis: the farming year*. Chapters 7 and 13. Chatto and Windus, London.

DUCKHAM, A. N. (1966). *The role of agricultural meteorology in capital investment decisions in farming*. Miscellaneous Studies No 2, Dept of Agriculture, University of Reading.

DUCKHAM, A. N. (1967). Weather and farm management decisions, in Taylor, J. A. (Ed) *Weather and Agriculture*. Pergamon Press, Oxford and New York.

DUCKHAM, A. N. and MASEFIELD, G. B. (1970). *Farming systems of the world 1970*, Chapters 1.2, 1.4, and 3.1. Chatto and Windus, London.

HMSO (1964). *The farmers' weather*. MAFF Bull. No 165 (2nd edition).

HOGG, W. H. (1964). Meteorology and agriculture. *Weather*, **19**, No 2, 34–43.

IVINS, J. D. (Ed.) (1959). *The measurement of grassland productivity*. Butterworth Scientific Publications, London.

LANGRIDGE, J. and McWILLIAM, J. R. (1967). Heat response of higher plants, in Rose, A. H. (Ed) *Thermobiology*. Academic Press, London.

MANNING, H. L. (1956). The statistical assessment of rainfall probability and its application to Uganda agriculture. *Proc. Roy. Soc. B*, **144**, 460.

MAUNDER, W. J. (1966). Climatic variation and agricultural production in New Zealand. *New Zeal. Geogr.*, **22**, 55–69.

MAUNDER, W. J. (1968). Agroclimatological relations: a review and a New Zealand contribution. *Canad. Geographer*, **12**, 73–84.

MAUNDER, W. J. (1968). Effect of significant climatic factors on agricultural production and incomes: a New Zealand example. *Mthly Weather Rev.*, **96**, 39–46.

MILTHORPE, F. L. and IVINS, J. D. (Eds) (1966). *The growth of cereals and grasses.* Butterworth, London.

SLATYER, R. O. (1960). *Agricultural climatology of the Yass valley.* Technical Paper No. 6, Division of Land Research and Regional Survey, CSIRO, Australia.

SMITH, L. P. (1958). *Farming weather* Nelson, London.

SMITH, L. P. (1967). Meteorology applied to agriculture. *WMO Bull.*, **16**, 190–4.

TAYLOR, J. A. (Ed) (1958-72). *Annual Symposia Series in Agricultural, Bio- and Applied Meteorology, Nos I-XV* (inclusive). Department of Geography, University College of Wales, Aberystwyth.

THOMAS, W. L. and EYRE, P. W. (1951). *Early potatoes.* Faber & Faber, London.

WADSWORTH, R. M. (Ed) (1968). *The measurement of environmental factors in terrestrial ecology.* Blackwell Scientific Publications, Oxford.

WAREING, P. F. and COOPER, J. P. (Eds) (1971). *Potential crop production.* Heinemann, London.

WEBSTER, C. C. and WILSON, P. N. (1966). *Agriculture in the tropics* (Chapter 1). Longmans, London.

WHYTE, R. O. (1960). *Crop production and environment.* Faber & Faber, London.

WILLIAMS, C. N. and BIDDISSCOMBE, E. F. (1965). Extension growth of grass tillers in the field. *Austr. J. Agric. Res.*, **16**, 14–22.

WILLIAMS, R. D. (1970). Tillering in grasses cut for conservation with special reference to perennial rye grass. *Herb. Abstr.*, 383–8.

WRIGLEY, G. (1969). *Tropical agriculture: the development of production.* Faber & Faber, London.

2 *Built Environment*

CHANDLER, T. J. (1965). *The climate of London.* Hutchinson, London.

INTERNATIONAL COUNCIL FOR BUILDING RESEARCH STUDIES AND DOCUMENTATION (1971). Knowledge of natural data and users' requirements, Topic 1. Natural Data: in *Proc. 5th Int. Cong. Paris, Research into Practice, the challenge of application.* CIB, Rotterdam, 9–48.

INTERNATIONAL COUNCIL FOR BUILDING RESEARCH STUDIES AND DOCUMENTATION (1972). Survey of available and required meteorological information for architecture and building. (CIB Working Commission W4A) CIB Rotterdam.

LANDSBERG, H. (1956). The climate of towns, in W. L. Thomas (Ed) *Man's role in changing the face of the earth.* University of Chicago Press, p 584.

PAGE, J. K. (1971). The meteorological loading of structures. *Building Sci.*, **6, (1)**, 17–23.

PAGE, J. K. (1971). Weather as a factor in building design and construction, in *Progress in Construction Science and Technology*, 65–102. Edited by R. A. Burgess *et al.*, MIP Press Ltd, Aylesbury.

ROYAL SOCIETY, LONDON (1971). Architectural Aero-dynamics. *Phil. Trans. Roy. Soc. Series A*, **269**, Pt. **1199**, 321–554.

SHELLARD, H. C. (1967). Wind records and their application to structural design. *Met. Mag.*, **96**, 235–43.

WORLD METEOROLOGICAL ORGANISATION (1970). *Urban climates*, Technical Note No 108, WMO No 254, TP 141, WMO, Geneva.

WORLD METEOROLOGICAL ORGANISATION (1970). *Building climatology*, Technical Note No 255, TP 142, WMO, Geneva.

3 *Climatic Change*

CURRY, L. (1962). Climatic change as a random series. *Ann. Ass. Am. Geogrs.*, **52**, 21–31.

LAMB, H. H. (1966). *The changing climate: selected papers*. Methuen, London.

MANLEY, G. (1961). The range and variation of the British climate. *Geog. J.*, **117**, 43–65.

MANLEY, G. (1964). The evolution of the climatic environment, in Watson, J. W. and Sissons, J. B. *The British Isles: a systematic geography*. Nelson, Ch. 9, p. 152–76.

TAYLOR, J. A. (1965a). Climatic change as related to altitudinal thresholds and soil variables, in Johnson, C. G. and Smith, L. P. (Eds) *Biological significance of climatic changes in Britain*. Academic Press, London and New York, 37–49.

TAYLOR, J. A. (1965b). Climatic change and Welsh agricultural development, in Taylor, J. A. (Ed) *Climatic change with special reference to Wales and its Agriculture*. Memo No 8, Geog Dept, UCW, Aberstwyth.

TAYLOR, J. A. (Ed). (1965c). *Climatic change with special reference to Wales and its agriculture*. Memo No 8, Geog Dept, UCW, Aberystwyth.

4 *Economic*

ACKERMAN, E. A. (1966). Economic analysis of weather: an ideal weather pattern model, in Sewell, W. R. D. (Ed) *Human dimensions of weather modification*. Univ of Chicago, Dept of Geog, Res Paper No 105, 61–76.

CZELNAI, R. *et al.* (1970). On the economic efficiency of meteorological activities, **Idojaras, 74,** 484–96 (in English). Meteorological Service of the Hungarian Peoples Republic, Budapest.

McQUIGG, J. D. and DOLL, J. P. (1961). *Weather variability and economic analysis*. Missouri College of Agriculture Res Bull 771, Columbia, USA.

TAYLOR, J. A. (Ed) (1970). *Weather economics*. Pergamon Press, Oxford.

TAYLOR, J. A. (1971). Curbing the cost of bad weather. *New Scientist and Science Journal*, 3 June 1971, 560–3.

THOMPSON, J. C. and BRIER, C. L. (1955). The economic utility of weather forecasts. *Mthly. Weather Rev.*, **11**, No 83, 249–54.

WORLD METEOROLOGICAL ORGANISATION (1967). *Assessing the economic value of a national meteorological service.* World Weather Watch Planning Report No 17, WMO, Geneva.

WORLD METEOROLOGICAL ORGANISATION (1968). *The economic benefits of national meteorological services.* World Weather Watch Planning Report No 27, WMO, Geneva.

5 *General*

BOER, W. (1964). *Technische Meteorologie.* Leipzig.

DRUCKER, P. F. (1964). *Managing for results.* Heinemann, London.

DRUCKER, P. F. (1967). *The effective executive.* Heinemann, London.

DRUCKER, P. F. (1969). *The practice of management.* Heinemann, London. (First edition 1954, Harper, New York; 1955, Heinemann, London.)

LAMB, H. H. (1964). *The English climate* (new edition). EUP, London.

MANN, R. E. (1966). *Descriptive micrometeorology.* Academic Press, London.

MAUNDER, W. J. (1970). *The value of weather.* Methuen, London.

MAUNDER, W. J. and WHITMORE, A. D. (1969). The value of weather: challenge of assessment. *Austral. Geogr.* **II**, 22–8.

MCDOWELL, R. E. (1968). Climate versus man and animals. *Nature, Lond.*, **218**, 18 May 1968, pp 141 *et seq.*

RAPP, R. R. and HUSCHKE, R. E. (1964). *Weather information: its uses, actual and potential.* Memo RM. 4083-USWB, The Rand Corporation, Santa Monica, California.

ROONEY, J. F. (1966). *The urban snow hazard: an analysis of the disruptive impact of snowfall at ten cities in the Central and Western USA.* Unpublished PhD thesis, Clark Univ, Worcester, Mass, USA.

SELLERS, W. D. (1965). *Physical climatology.* Univ of Chicago Press, Chicago and London.

SEWELL, W. R. D. (1968). *Human dimensions of the atmosphere.* National Science Foundation, Washington DC (contains classified bibliographies).

SEWELL, W. R. D., KATES, R. W., and PHILLIPS, L. P. (1968). Human response to weather and climate: a geographical contribution. *Geog. Rev.*, **58**(3), 262–80.

TOWNSEND, R. (1970). *Up the organisation.* Michael Joseph, London.

6 *Hydrology*

CRAINE, L. E. (1970). *Water management innovations in England.* Johns Hopkins Press, Baltimore.

CRAWFORD, N. H. and LINSEY, R. K. (1966). *Digital simulation in hydrology: Stanford watershed model IV.* Tech Rep 39, Stanford Univ, Dept of Civil Eng.

HEWINGS, J. M. (1968). Water quality and the hazard to health: placarding public beaches. Univ of Toronto, *Natural Hazard Research Working Paper*, 3, 27 pp.

INSTITUTION OF CIVIL ENGINEERS (1966). *River flood hydrology*. Symposium.

INSTITUTION OF WATER ENGINEERS (1969). *River-flow management*. Proc. Symp. at Newcastle upon Tyne, September, 1966.

ISAAC, P. C. G. (Ed) (1967). *River management*. Proc. Symp. at Newcastle upon Tyne, September, 1966.

METEOROLOGICAL OFFICE (UK) (1968). *Rain over the catchment area of the Trent in relation to flooding at Nottingham*. MO Report,

THORN, R. B. (Ed) (1966). *River engineering and water conservation works*. Butterworth, London.

7 Industrial

BICKERT, C. VON E. and BROWNE, T. D. (1966). Perception of the effects of weather on manufacturing: a study of five firms, in Sewell, WRD (Ed) *Human Dimensions of Weather Modification*. Univ of Chicago, Dept of Geog Res Paper No 105, 307–22.

COLLINS, G. F. (1956). A severe weather service for industry. *Bull. Am. Met. Soc.*, 47, No 10, 514–17.

McQUIGG, J. D. and THOMPSON, R. G. (1966). Economic value of improved methods of translating weather information into operational terms. *Mthly Weather Rev.*, 94, 2, 83–7.

RUSSO, J. A. (1966). The economic impact of weather on the construction industry of the United States. *Bull. Am. Met. Soc.*, 47, 967–72.

8 Pollution

DE GROOT, I. and SAMUELS, S. W. (1963). *People and air pollution: a study of attitude in Buffalo, New York*. Dept. of Health, Education & Welfare, Washington DC.

LYCAN, D. R. and SEWELL, W. R. D. (1967). *Water and air pollution as components of the urban environment of Victoria*. Paper presented to the Annual Meeting of the British Columbia regional division of the Canadian Association of Geographers, Vancouver, BC. 5 February 1967.

MEDALIA, N. Z. (1964). Air pollution as a socio-environmental health problem: a survey report. *J. Health and Human Behaviour*, 5, 154–65.

MEDALIA, N. Z. (1965). Community perception of air quality: an opinion survey in Clarkston, Washington. *Public Health Survey Publication*, 999-AP-10.

OGDEN, D. C. (1966). Economic analysis of air pollution. *Land Econ.*, 42, 137–47.

SCHUSKY, J. (1966). Public awareness and concern with air pollution in the St Louis metropolitan area. *J. Air Pollution Control Assoc.*, 16, 72–6.

TYSON, P. D. (1963). Some climatic factors affecting atmospheric pollution in South Africa. *South African Geog. J.*, 45, 44–54.

9 *Social, including Tourism*
BATES, M. (1966). The role of weather in human behaviour, in Sewell, W. R. D. (Ed) *Human dimensions of weather modification*, Univ of Chicago, Dept of Geog Res Paper No 105, 393–407.
BURTON, I., KATES, R. W., and WHITE, G. P. (1968). *The human ecology of extreme geophysical events*. Natural Hazard Research Working Paper, No 1, Toronto: Univ of Toronto, Dept of Geog, 1968.
BURTON, T. L. (1970) *Recreation research and planning*. George Allen & Unwin, London.
CLAWSON, M. (1966). The influence of weather on outdoor recreation, in Sewell, W. R. D. (Ed), *Human dimensions of weather modification*. Res Paper No 105, Dept of Geog, Univ of Chicago, 183–93.
CLAWSON, M., and KNETSCH, J. L. (1966). *Economics of outdoor recreation*, Johns Hopkins Press, Baltimore, Md.
COVERT, R. P., GOLDHAMMER, M. M., and LEWIS, G. F. (1967) An estimation of the effects of precipitation of scheduling of extended outdoor activities. *J. Appl. Met.* 6(4), 683–7.
CURRY, L. (1952). Climate and economic life: a new approach with examples from the USA. *Geogr. Rev.*, 42, 367–83.
HALLANGER, N. L. (1963). The business of weather: its potentials and uses. *Bulletin of the American Meteorological Society*, 44, No 2.
HEURTIER, R. (1968). Essai de climatologie touristique synoptique de l'Europe occidentale et Méditerranée pendant la saison d'été. *La Meteorologie*, 71–107.
HOLFORD, I. (1967) Planning your days according to the weather. *Weather*, 22, No. 4, 132–3.
MORRIS, E. A. (1966). Institutional adjustment to an emerging technology: legal aspects of weather modification. In Sewell, W.R.D. (Ed), *Human dimensions of weather modification*, Research Paper No 105, Department of Geography, University of Chicago, 279–88.
PATMORE, J. A. (1970). *Land and leisure*. David & Charles, Newton Abbot, England.
PERRY, A. H. (1972). Weather, climate and tourism. *Weather*, 27, No 5, 199–203.
SHAFER, E. L. & THOMPSON, R. C. (1968). Models that describe use of Adirondack campgrounds. *Forest Science*, 14, No 4, 383–91.
STAFF, J. M. (1961). Meteorology and the community. *Q. J. Roy. Met. Soc.*, 87, 465–71.
TERJUNG, W. H. (1968). Some thoughts on recreation geography in Alaska from a physio-climatic viewpoint. *The Californian Geographer*, 9, 27–39.
US DEPT OF COMMERCE WEATHER BUREAU (1964). *The national research effort on improved weather description and prediction for social and economic purposes*. US Weather Bur, Washington DC.

10 *Transportation*

ANON. (1970). Way shown to monitor bridge icing. *Rural and Urban Roads*, No 5, 42–3.

BIRNIE, C. and MEYER, W. E. (1970). *Prediction of preferential icing conditions on highway bridges.* Highway Research Board Special Reports, No 115, Highway Research Board National Academy of Sciences, Washington DC, USA, 27–35.

CIEMOCHOWSKI, M. F. (1968). Simple detector predicts formation of ice on roads. *The SAE Journal*, **76**, No 8, 60–1.

CIEMOCHOWSKI, M. F. (1969). *A detection system for frost; snow and ice on bridges and highways.* Highway Research Record No 298, Highway Research Board, National Academy of Sciences, Washington DC, USA.

INQUE, M. *et al.* (1970). *Ice detection, prediction and warning system on highways.* Highway Research Board Special Reports, No 115, Highway Research Board, National Academy of Sciences, Washington DC, USA, 17–26.

MATHEWS, A. (1965). 'Spot' icing major culprit in winter death, injury toll. *Traffic Eng.*, **35**, No 5, 10–12.

SCHEIDER, T. R. (1969). Auswertung mikroklimatischer Messunger an Strassen-korpern (Evaluation of microclimatic measurements undertaken on road structures). *Strasse und Verkehr*, 53–9.

11 *Weather and Climatic Hazards*

BURTON, I. *et al.* (1969). *Human ecology of coastal flood hazard in Megalopolis.* Univ of Chicago, Dept of Geog, Res Paper No 115.

EDWARDS, R. S. (1969). Economic measurement of weather hazards. *Weather*, **24**, 2, 70–3.

ENVIRONMENTAL SCIENCE SERVICES ADMINISTRATION (1965). A proposed nation-wide natural disaster warning system. *ESSA*, Washington DC, 25–37.

ERICKSEN, N. J. (1971). Human adjustment to floods in New Zealand. *NZG*, **27**, 105–29.

GOLANT, S. and BURTON, I. (1968). *Avoidance-response to the risk environment.* Univ of Toronto, Geog Dept Natural Hazard Working Paper 6.

GOLANT, S. and BURTON, I. (1970). A semantic differential experiment in the interpretation and grouping of environmental hazards. *Geogr. Analysis*, **2**, 120–34.

HELBUSH, R. E. (1968). Linear programming applied to operational decision-making in weather risk situations. *Mthly Weather Rev.*, **96**, 876–82.

INSTITUTION OF CIVIL ENGINEERS (1967). Flood studies for the United Kingdom. Report: Committee on Floods in the UK.

KATES, R. W. (1971). Natural hazard in human ecological perspective: hypotheses and models. *Econ. Geog.*, **47**(3), 438–51.

RALPH, E. C., GOODWILLIE, S., BURTON, I., and SCHULTE, P. (1968). *Annotated bibliography of snow and ice problems.* Univ of Toronto, Geogr Dept, Natural Hazard Res Paper, **2,**

RUSSELL, C. J. (1969). *Losses from natural hazards.* Univ of Toronto Geog Dept, Natural Hazard Research Working Paper, **10f.**

THORNTHWAITE, C. W. (1965). *The shores of Megalopolis: coastal occupance and human adjustment to flood hazard.* C. W. Thornthwaite Associates, Route 1, Elmer, New Jersey 08318, USA.

12 *Weather Forecasting*

BROOME, M. R. (1966). Weather forecasting and the contractor. *Weather,* **21,** 406–10.

MASON, B. J. (1970). Future developments in meteorology: an outlook to the year 2000. *Quart. J. Roy. Met. Soc.* **96,** 349–68.

THOMPSON, J. C. (1965). Advances in weather forecasting. *Weather,* **20,** No 3, 74–9.

WICKHAM, P. G. (1970). The practice of weather forecasting. HMSO, London.

13 *Weather Modification*

ACKERMAN, E. A. (1959). Weather modification and public policy, in Jarrett, H., (Ed), *Science and Resources.* Johns Hopkins Press, Baltimore, USA, 63–74.

FLEAGLE, R. C. (1969). *Weather modification: science and public policy* (new edition). Univ of Washington Press, London and Seattle.

KRICK, I. P. (1955). *A bibliography of weather modification and field operations using the combined facilities of I. P. Krick, American Institute of Aerological Research and Water Resources Development Corporation, Denver, Colorado, USA.* Am. Inst. Aero. Res., August 31.

MALONE, T. F. (1967). Weather modification: implications of the new horizons in research. *Science,* **156 (13777),** 897–901.

MORRIS, E. A. (1966). Institutional adjustment to an emerging technology: legal aspects of weather modification, in Sewell, W. R. D. (Ed). *Human dimensions of weather modification.* Research Paper No 105, Dept of Geog, Univ of Chicago, 279–88.

PETTERSSON, S. (1964). Meteorological problems: weather modification and long-range forecasting. *Bull. Am. Met. Soc.,* **45,** No 1, 2–6.

SEWELL, W. R. *et al.* (1966). Perception of possibilities of weather modification and attitudes towards government involvement, in Sewell, W. R. D. (Ed) *Human dimensions of weather modification.* Univ of Chicago, Dept of Geog Res Paper No 105, 329–46.

SEWELL, W. R. D. (Ed) (1969). *Human dimensions of weather modification.* Univ of Chicago, Geog Res Paper No 105, 195–207.

Appendix: Meteorological Office Services— Weather advice to the community (reproduced by permission of the Director of the Meteorological Office)

Introduction

As the State Weather Service the Meteorological Office includes among its functions the provision of meteorological services required by government departments, public corporations, local authorities, industry and commerce, radio and television, the Press, and the general public.

To keep the public informed of current and expected weather in and around the United Kingdom, weather reports and forecasts are issued frequently through public information channels. Special needs not covered by these weather bulletins may be met by arrangement.

Records held by the Meteorological Office include returns and observation registers from a large number of meteorological stations at home, from British stations overseas, and from many ships which provide reports of weather at sea. These records may be consulted by anyone.

The Meteorological Office provides advisory services on all aspects of climatology, rainfall and evaporation, agricultural and marine meteorology, and on meteorological instruments.

Members of the public may use the National Meteorological Library which has a wide range of textbooks, technical journals and reports, collections of photographs and transparencies, and a very large quantity of published meteorological data from countries throughout the world.

Services are provided by the Headquarters of the Meteorological Office and by the National Meteorological Library at Bracknell, Berkshire, and by many offices throughout the United Kingdom. This appendix describes the services and how to obtain them.

Definitions

A *weather report* is a statement of observed weather that existed at a specified time and place.

A *weather forecast* is a statement of weather expected to exist for a specified period of time in a stated locality.

A *weather warning* is a notification of the expected onset of specified weather (eg snow, thunderstorms, frost, gales, fog) in a stated locality.

Services provided through public information channels

Short-period forecasts, mainly designed to give general guidance, are broadcast on radio and television, are available in many areas on the Post Office's automatic telephone weather service and are printed in newspapers.

Radio. Weather forecasts are broadcast regularly in the various BBC services. Details are given in the *Radio Times.*

National forecasts are transmitted regularly on Radio 1 and 2, most frequently in the early morning. Other Radio 2 transmissions may be interrupted for 'flash' warnings of weather liable to cause great inconvenience to large numbers of the population and for warnings of fog and strong crosswinds on motorways. Radio 2 also transmits gale warnings and forecasts for shipping.

Radio 3 transmits brief national forecasts three or four times daily.

The main weather bulletins on Radio 4 normally include a statement of the general weather situation, a national forecast for some 18 to 24 hours ahead, an outlook for a further 24 to 48 hours and a more detailed regional forecast. BBC Regions and Radio Stations are shown on Fig. A.1. The bulletins also contain weather warnings applicable at the time. There are daily weather bulletins for farmers and for mariners in coastal waters.

Local radio stations broadcast weather forecasts for the local area.

Most news bulletins include a summary of the national forecast.

Any of the forecasts supplied to the BBC may be copied and displayed for public information but if this is done the times of broadcasting of the forecast and the period of validity must be stated as these are essential parts of the forecast.

Television. On BBC a forecaster of the Meteorological Office displays charts several times each day showing the current weather situation and the situation expected the following day. He describes the weather expected during the coming twenty-four hours in general terms for the country as a whole and gives a brief outlook for a further day or two. A weekly bulletin designed for farmers is broadcast each Sunday afternoon. Times of all these transmissions and the times at which more detailed regional forecasts are given in Regional programmes are published in the *Radio Times.*

The independent television companies broadcast weather bulletins based on information supplied by the Meteorological Office. The times of broadcast are given in *TV Times.*

The automatic telephone weather service. Local area weather forecasts for many parts of the country are prepared by the Meteorological Office and recorded by the Post Office. The area telephone directories give details of the services available in the area and their telephone numbers. There is no charge apart from the cost of the telephone call. The service is being expanded and notices of its introduction in new areas are displayed in local post offices.

Fig. A.1 BBC Regions and Radio Stations 1972

For foreign tourists in the London area, recorded local weather forecasts are available in French by dialling 246 8043, in German on 246 8045, in Spanish on 246 8047 and in Italian on 246 8049.

The Press. Weather forecasts for all districts of the United Kingdom and for Irish Sea and English Channel crossings are issued regularly to newspapers throughout the country. The districts referred to in these forecasts are shown on the map on p 239 and their boundaries are defined on pp 238–41.

The text of all forecasts issued to the Press is liable to editorial changes by the Press and the period of validity is not always clearly indicated in the printed forecast

Special Services

Weather forecasts on public information channels do not meet all needs for weather advice. To meet more specialised or detailed needs the Meteorological Office provides a number of special services described below. These fall into three main categories.

The first covers weather reports, forecasts, and warnings to meet the special needs of the inquirer. The second type of service is concerned with the study of records of past weather and the preparation and interpretation of summaries and statistical analyses of the weather in a particular locality. It allows for consultation with a meteorologist and for special investigations to help assess the effect of weather on a particular activity and to assist planning and development. The third service is an educational one, providing for the loan of photographs, transparencies, and films on meteorology and the work of the Office.

Service 1A. The issue of weather reports by prior arrangement.

Service 1B. A standing arrangement for notification of the reported occurrence of specified weather in a stated locality anywhere in the United Kingdom.

Service 1C. An arrangement for the issue of weather forecasts, singly or regularly, for any locality in the United Kingdom. Forecasts are issued for any period up to 24 hours from the time of issue. When requested, an outlook for a further 24 hours is added. In certain weather situations it may be possible to extend the outlook.

Service 1D. A standing arrangement for notification of the expected occurrence of specified weather in a stated locality, eg warnings of snow, thunderstorms, frost, gales, fog, anywhere in the United Kingdom.

Service 1E. Notification by telephone or telegram on all occasions during the period 1 May to 31 October when fine weather spells are expected anywhere in the United Kingdom. Cancellation messages are sent when fine spells are expected to end. A fine spell is defined for the purpose of this service as a period of at least 72 hours without measurable rain.

Service 1F. Warnings of certain road dangers due to weather (snow or icy surfaces) during the period 1 October to 31 May.

Service 1G. Notification by telegram every Friday of expected weekend temperatures for any area in the United Kingdom during the period first Friday in November to third Friday in April. The telegrams give the forecast mean and minimum temperatures for the periods noon Saturday to noon Sunday and noon Sunday to noon Monday.

Service 1H. Monthly Weather Survey and Prospects. A statement of the weather prospects in Britain for the next 30 days is published at the beginning and in the middle of each month. The publication at the beginning of the month also describes the previous month's weather and the climatology characteristic of the coming month.

Service 2A. Personal consultation with a meteorologist at certain offices.

Service 2B. Outside attendance by a professional member of the staff, eg to appear at a public inquiry or in a court in connection with a civil action, or to act as a consultant.

Service 2C. Investigation, computation or other professional work as required. Statements provided under this service may be certified if desired. The certified statements of the Meteorological Office are usually accepted by Courts of Law without personal evidence.

Service 2D. The supply of copies of data, autographic records, etc.

Service 2E. The extraction of data by visitors to the Bracknell, Edinburgh and Belfast Offices, subject to normal library and archive rules.

Service 3. The loan of photographs, transparencies, and films illustrating such subjects as the work of the Meteorological Office, clouds and other weather phenomena and meteorological instruments.

Charges for all the services described above are obtainable from the Meteorological Office. Additional fees may be charged under copyright regulations and suitable acknowledgements may be required if the material is used for publication.

The arrangements for coastwise shipping and fishing vessels are given in *Meteorological Office Leaflet No 3.* Certain facilities are also available for farmers and growers through some of the regional headquarters of the National Agricultural Advisory Service; details are obtainable from the local Agricultural Advisory Officer.

Other services to meet special needs, eg of weather-sensitive activities, may be provided on request at rates to be negotiated separately.

For special services apply to:

Services 1A to 1D. The Senior Meteorological Officer at any office listed on p 237.
    The applicant should give:
        The period and place or area to be covered.
        The time at which the information is desired.
        The Telex or telephone number to which the information is to be sent.

It is advisable to give the purpose for which a weather report or forecast is required, specifying any weather features of special interest (eg wind, rain, temperature). This information will often help in the preparation of the report or forecast.

Services 1E to 1G
  The Director-General
  Meteorological Office, Met. O. 7a
  London Road
  Bracknell, Berks RG12 2SZ   Bracknell 20242

Service 1H
  The Secretary
  Meteorological Office, Met. O. 10a/DWR
  London Road
  Bracknell, Berks RG12 2SZ   Bracknell 20242

Services 2A to 2E
For England and Wales:
  The Director-General
  Meteorological Office, Met. O. 3
  London Road
  Bracknell, Berks RG12 2SZ   Bracknell 20242

For Scotland:
  The Superintendent
  Meteorological Office
  26 Palmerston Place
  Edinburgh, EH12 5AN   Caledonian 6561

For Northern Ireland:
  The Senior Meteorological Officer
  Tyrone House
  Ormeau Avenue
  Belfast 2
  BT2 8HH   Belfast 28457

Service 3. Inquiries about photographs and transparencies should be addressed to:
  The National Meteorological Library
  London Road
  Bracknell, Berks RG12 2SZ   Bracknell 20242

For films apply to:
  Services Kinema Corporation
  Chalfont Grove
  Narcot Lane
  Chalfont St Peter
  Gerrards Cross, Bucks   Chalfont St Giles 4111

Alternative arrangements for all services. Any special service may be arranged by applying as in *Services 1E to 1G* above.

Forecast districts/England and Wales

1 *Greater London*
2 *South-east England*
   Kent
   Surrey
   Sussex
3 *East Anglia*
   Norfolk
   Suffolk
   Cambridgeshire
   Essex
4 *Central Southern England*
   Berkshire
   Hampshire
   Wiltshire
   Dorset
   Isle of Wight
5 *East Midlands*
   Leicestershire
   Rutland
   Northamptonshire
   Huntingdonshire
   Bedfordshire
   Hertfordshire
   Oxfordshire
   Buckinghamshire
6 *East England*
   Lincolnshire
   Nottinghamshire
   Yorkshire (East Riding)
7 *West Midlands*
   Staffordshire
   Shropshire
   Worcestershire
   Warwickshire
   Herefordshire
   Gloucestershire

8 *Channel Islands*
9 *South-west England*
   Somerset
   Devonshire
   Cornwall
   Isles of Scilly
10 *South Wales*
   Cardiganshire
   Radnorshire
   Pembrokeshire
   Carmarthenshire
   Breconshire
   Glamorgan
   Monmouthshire
11 *North Wales*
   Anglesey
   Caernarvonshire
   Denbighshire
   Flintshire
   Merionethshire
   Montgomeryshire
12 *North-west England*
   Lancashire
   Cheshire
13 *Lake District*
   Cumberland
   Westmorland
14 *Isle of Man*
15 *Central Northern England*
   Yorkshire (West Riding)
   Derbyshire
16 *North-east England*
   Northumberland
   Durham
   Yorkshire (North Riding)

Forecast districts/Scotland

17 *Borders*
   The counties of Berwick, Roxburgh, Peebles, Selkirk, and that part of Dumfries-shire east of Annandale.

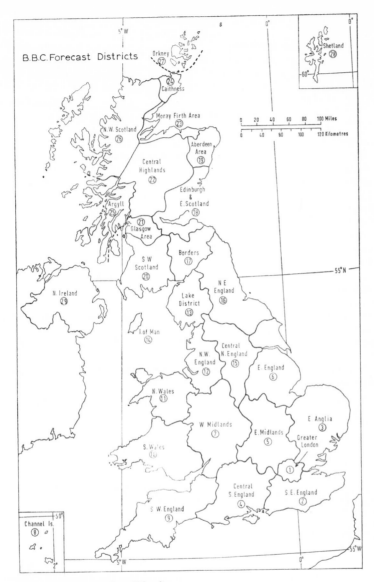

Fig. A.2 **Forecasting Districts**

18 *Edinburgh and East Scotland*
The part of Kincardine southward of a line running west from Stonehaven. Angus south-east of a line Edzell to Alyth. Perthshire south and east of a line through Alyth, Dunkeld, Methven and due south from Methven to the borders of Kinross. Fife, Kinross-shire and Clackmannanshire, East Stirlingshire (East of the Campsie Fells), West Lothian, Midlothian, and East Lothian.

19 *Aberdeen Area*
The whole of Aberdeenshire except that part lying west of a line from Mount Battock to the Buck.
The part of Kincardine northward of a line running west from Stonehaven.

20 *South-west Scotland*
Dumfriesshire west of and including Annandale, Kirkcudbrightshire, Wigtownshire, and Ayrshire.
Lanarkshire except that part of the Glasgow area as defined below.
Bute and Arran.

21 *Glasgow Area*
The whole of Renfrewshire.
The industrial part of Lanarkshire which includes Glasgow, Airdrie, Motherwell, Wishaw, Carluke, Lanark, Hamilton, and East Kilbride.
The country of Stirling south-west of the Campsie Fells including Kilsyth and Drymen.
That part of Dunbartonshire which includes Dumbarton and Alexandria and all places south-west of a line through Dumbarton, Alexandria, and Drymen.

Note: There is a part of Dunbartonshire between Lanarkshire and Stirlingshire. This is also included in the Glasgow area.

22 *Central Highlands*
Parts of Inverness-shire comprising the Great Glen and the area east of the Great Glen (except the area within 10 miles of the Burgh of Inverness).
Aberdeenshire west of a line from Mount Battock to the Buck.
The highland parts of Nairn, Moray, and Banff.
Angus north-west of Strathmore.
Perthshire north and west of a line through Alyth, Dunkeld, Methven and due south from Methven to the boundary of Kinross, including Alyth, Dunkeld, and Methven and thence along the Ochils.
Parts of the countries of Dunbartonshire and Stirling north of a line through Dumbarton, Alexandria, Drymen, the Campsie Fells, and Stirling, but excluding these places.

23 *Moray Firth Area*
Bonar Bridge and the eastern coastal strip of Sutherland. (This does not include Lairg.)
Ross and Cromarty east of Ben Wyvis and Ben Tharsuinn. The Burgh of Inverness and the county of Inverness within 10 miles of the county town.
The lowlands of the counties of Nairn, Moray, and Banff.

24 *Caithness*

25 *Argyll*
The county of Argyll.

26 *North-west Scotland*
The whole of Sutherland except Bonar Bridge and eastern coastal district (Lairg is included in this district).
The whole of Ross except that part east of Ben Wyvis and Ben Tharsuinn.
Inverness-shire west of the Great Glen (except the area within 10 miles of the Burgh of Inverness). The island parts of Inverness-shire and Ross.

27 *Orkney*

28 *Shetland*

Forecast districts/Northern Ireland
29 *Northern Ireland*
Londonderry, Antrim, Tyrone, Fermanagh, Armagh, Down.

Meteorological offices serving the community

England and Wales

*	Abingdon, Berks	Abingdon 1408
	Bawtry, near Doncaster	Bawtry 474
	Birmingham Airport	021–743 4747
†	Boscombe Down, Wilts	Amesbury 3331, ext. 2131
†	Chivenor, Devon	Barnstaple 3722
‡	Glamorgan Airport	Rhoose 343
	Gloucester	Gloucester 23122
	Honington, Suffolk	Bury St Edmunds 5026
†	Linton-on-Ouse, Yorks	Linton-on-Ouse 261
	Liverpool Airport	051–427 4666
	London Weather Centre	01–836 4311
	Lyneham, Wilts	Bradenstoke 283
	Manby, Lincs	Louth 2145
	Manchester Weather Centre	061–832 6701

† Marham, Norfolk	Narborough 398
Newcastle Weather Centre	Newcastle upon Tyne 26453
† Oakington, near Cambridge	Willingham 555
Plymouth, Devon	Plymouth 42534
Preston, Lancs	Preston 52628
* St Mawgan, Cornwall	Newquay 2224
† Shawbury, near Shrewsbury	Shawbury 335
‡ Southampton Weather Centre	Southampton 28844
* Thorney Island, Hants	Emsworth 2355
Upavon, Wilts	Upavon 286
† Valley, Anglesey	Holyhead 2288
‡ Watnall, Nottingham	Nottingham 55155
† Wittering, Northants	Stamford 4802
* Wyton, Hunts	Huntingdon 2451, ext. 458

Scotland

‡ Aberdeen Airport	Dyce 334
‡ Edinburgh Airport	031–334 7777
Glasgow Weather Centre	041–248 3451
Kinloss, Morayshire	Forres 2161, ext. 116
‡ Kirkwall Airport, Orkney	Kirkwall 2421, ext. 27
† Leuchars, Fife	Leuchars 224
Pitreavie, near Dunfermline	Inverkeithing 2566
Prestwick Airport	Prestwick 78475

Northern Ireland

Belfast Airport	Crumlin 339

Offices are open 24 hours a day, seven days a week except as indicated:
  * Closed Friday evening, Saturday, and Sunday.
  † Open office hours (Monday to Friday) only.
  ‡ Open 24 hours per day, but there may be short delays in service in the late evening and during the night.

On request any of these offices can supply the latest forecast for any area of the United Kingdom free of charge.

The Weather Centres in London, Manchester, Newcastle, Southampton, and Glasgow are open for personal inquiries all day from Monday to Friday.

The information in this Appendix is liable to constant amendment and the Meteorological Office and/or the communication media should be consulted as required.

Author Index

NOTE: Authors listed alphabetically at the end of each chapter and in the preceding Selected Classified Bibliography have been excluded from this index. References to illustrations (Figures) are in italics; references to Tables are prefixed with the letter 'T'. All references per author have been placed in chronological order within the sequences of page references.

243

Subject Index